THE MEDITERRANEAN WAS A DESERT

The Mediterranean Was a Desert

A VOYAGE OF THE
GLOMAR CHALLENGER

Kenneth J. Hsü

Princeton University Press
Princeton, New Jersey

Library of Congress Cataloging in Publication Data will be found
on the last printed page of this book

First Princeton Paperback printing, 1987

This book has been composed in Linotron Trump

Clothbound editions of Princeton University Press books are printed
on acid-free paper, and binding materials are chosen for strength
and durability. Paperbacks, while satisfactory for personal collections,
are not usually suitable for library rebinding.

Printed in the United States of America
by Princeton University Press, Princeton, New Jersey

To Christine, Elisabeth, Martin, Andreas, Peter,
and all those who stayed home

CONTENTS

—O—

CONTENTS

LIST OF ILLUSTRATIONS

PREFACE

—Θ—

THAT the Mediterranean could have dried up was a fascinating discovery. The story received far more press coverage than other more important but less dramatic scientific achievements. Bill Ryan and I were co-chief scientists on the deep-sea drilling cruise to the Mediterranean in 1970 when the drama unfolded. We were assisted by an expert team of paleontologists and sedimentologists, who provided invaluable advice in matters pertaining to their specialties, and by a hard-working drilling crew, which formed the backbone for the drilling operations. Our discovery was made possible by those who sent the drilling ship *Glomar Challenger* to the Mediterranean: by the Joint Oceanographic Institutions Deep Earth Sampling (JOIDES) group as the planner, the Deep Sea Drilling Project (DSDP) as the operator, the National Science Foundation as the sponsor, the United States Congress as the fund-raiser, and the witty pioneers of the now defunct "American Miscellaneous Society" who were imaginative enough to conceive the idea of deep ocean drilling. Team work, obviously, was an essential element in our success.

The term "drilling cruise" has been translated *Bohrungskampagne* by some German-speaking journalists. And indeed, *Kampagne* or "campaign," is perhaps a better description of our voyage than "cruise." We did not go on a cruise. We went to war against Nature, and we almost lost all the battles. After six weeks of continued frustration, one of the crew members formulated "Charlie's Law": "There are one hundred ways to do the wrong thing, and we are trying them all!" Our operations logbook reads in part like a book of unmitigated disaster. We were repeatedly prevented by a hard rock formation from reaching our objectives. But our troubles came to a happy resolution; Nature did yield its secret, though grudgingly, and we had our story.

For two months there were sixty-nine of us on board the *Glomar Challenger*. We came from different walks of life, but we all worked for big science. There were roughnecks, sailors, the operations manager, the captain, the marine technicians. They came to do a difficult job and took pride in doing it well. But they did not ask for relevance. Then there were the shipboard scientists. The few of us in the "decoding" section were at times quite emotionally involved in our task, so there were disagreements, compromises, complaints, and misunderstandings. But traces of humor occasionally seeped through. I jotted down the first draft of this book in the driller's shack of the drilling vessel during the heat of the "battle," recording in it our momentary joys, angers, and frustrated ambitions. Our sensibilities must have seemed rather petty to the drilling crew. Indeed, in the years since then, I have grown to share their conventional wisdom; our difficulties were not all that important, and life on the *Challenger* remains only a fond memory. We were all very human, very trivial—a big, happy, quarreling family in a small oasis, isolated from reality, and far from the madding crowd.

Technical terms, abbreviations, and acronyms are often unavoidable in the text, but I have tried to give an explanation the first time such an expression is used if the flow of the text is not interrupted. Otherwise the reader should refer to the glossary at the end of the book.

I have revised my manuscript for publication with the encouragement of Ed Tenner, the science editor at Princeton University Press. I am also indebted to Xavier Le Pichon for editorial advice and to Carolina Hartendorf for secretarial service. I have enjoyed working with Tam Curry, an excellent copyeditor with Princeton University Press. I acknowledge with thanks the persons and organizations who have given me permission to reproduce their photographs as illustrations in this book: Olivier Leenhardt (fig. 4); Marie Tharpe (fig. 1); American Geophysical Union (figs. 16 and 17); Geological Society of America (fig. 25); Deep Sea Drilling Project (figs. 12, 13, 14, 18, 19, 20, 21, 22, 24, 28, and 36). I would also like to thank Urs Gerber and Albert Uhr for their photographic work. Finally, I would like to express my high esteem for Bill Ryan; my text is testimonial to a friendship born at sea.

Zurich, June 1982   K. J. Hsü

THE MEDITERRANEAN WAS A DESERT

1. Schematic diagram showing the submarine topography of the Mediterranean seabed; prepared from bathymetry studies by Bruce Heezen, Marie Tharpe, and William B. F. Ryan of the Lamont-Doherty Geological Observatory. Photograph courtesy of Marie Tharpe.

PROLOGUE

—Θ—

It was in those days, before the ocean waters broke into the Mediterranean, that the swallows and a multitude of other birds acquired the habit of coming north, a habit that nowadays impels them to brave the passage of the perilous seas that flow over and hide the lost secrets of the ancient Mediterranean valleys.—H. G. Wells, *The Grisly Folk*

BILL RYAN and I had stayed up all night again. We were both very tired, not having had a good night's sleep since we left Lisbon ten days earlier. It was the morning of August 24, 1970, and the *Glomar Challenger* was positioned 180 kilometers off the Barcelona coast (fig. 2). Ryan was discouraged. He had invested some ten years of his professional life in studying the Mediterranean seabed with all manner of sophisticated geophysical gadgets. He knew that there was some very strange rock down there, reflecting back all acoustic signals sent down. Now we had a drilling ship and could bore through the deep-sea floor. A sample, or even a small chip, of the rock might resolve his puzzlement. Yet the answer had eluded us; we did not seem to be able to bring in any of the mystery rock. The night before, when we had finally thought we were right on top of our target horizon, the drill pipe had gotten stuck in the hole. When at last they raised the core barrel, they called me, but we could find nothing in it except sands and gravels.

Reading *A Tale of Two Cities* as a youth, I was puzzled

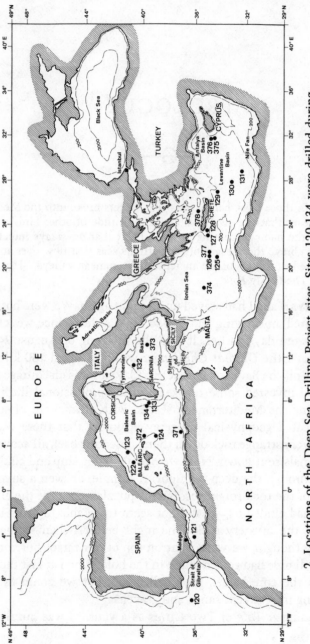

2. Locations of the Deep Sea Drilling Project sites. Sites 120-134 were drilled during the 1970 Leg 13 expedition, and Sites 371-378 were drilled during the 1975 Leg 42A expedition. Also shown are 200-meter and 2,000-meter contours.

by the strange behavior of the doctor who always took out his cobbler's tools and started pounding when he had to face unpleasant situations or crises. But working with Ryan made me aware of the therapeutic value of losing oneself in trivial chores. Ryan did not smile when he saw the gravels. He simply placed the bucket in the sink of the core lab and started to wash away the fine silts and clays. Then he sorted out the sands and pea-sized gravels, had them dried on a hot plate, and glued them in neat rows on a manila folder. I sat there watching in silence. As his collection grew, however, I began to take interest. The coarser grains, or the "pea gravels," measured some five to seven millimeters across. The shiny crystals were gypsum, a sulphate of calcium and an evaporite mineral, so called because it was an evaporative residue of seawater. Gypsum occurs today in muddy sediments on arid coasts, and it is present in older evaporite formations on land, but no one would expect to find gypsum in a deep-sea core, still less in a gravelly deposit.

Sands and gravels themselves are uncommon on the deep-sea floor, where the terrigenous sediments consist mainly of clays. Occasionally an avalanche of loose debris down a steep submarine slope might generate an underwater mud-flow, known to geologists as a turbidity current, and such currents might bring coarse debris from a sandy beach to an abyssal plain hundreds of kilometers away. But gypsum sands or gravels had never been reported on the deep-sea floor. Besides, we would expect to find gravels of other rock types derived from erosion on land if the gypsum had come from an older formation on the Spanish coast. There should be, first of all, quartz, which is the major constituent of all sands. Then there should be feldspar, fragments of granite, of rhyolite, of gneisses, schists, and other metamorphic rocks, and perhaps also clasts of quartzites, sandstones, shales, and carbonate rocks that had been laid down on continental shelves. We found none of these. Instead, we found a strange association of erosional debris: in addition to the gypsum, we could identify three other components that are rarely found in any gravel deposit, namely oceanic basalts, hardened oceanic oozes, and an unusual fauna of very tiny shells. All the debris seemed to have been derived from a seabed—more particularly, a desiccated seabed.

Geophysical studies had told us that a submarine volcano lay not far from the drill site, but the volcano could not have yielded pea gravels unless it was once exposed subaerially and subjected to stream erosion. Gravels of oceanic oozes were practically unheard of. Oozes, consisting of minute skeletons of marine organisms, usually remain soft in the deep-sea bottom, unless they have been buried to a great depth, or unless they have been exposed to air and dried up by sunlight. Even then, stream erosion and transport would be necessary to convert a seabed of hardened ocean oozes into a deposit of sands and gravels. Finally, the unusual fauna, as we were told by paleontologists on board, belonged to species living in coastal lagoons, except that they were "dwarfs." Our borehole had been spudded in 2,000 meters of water, however, and there could be no lagoon at such a depth unless the Mediterranean had once lost much of its water. Was it possible, then, that the Mediterranean had once been isolated from the Atlantic and had been changed into a desert?

I began to imagine Gibraltar's once being an isthmus that prevented water from flowing between the Atlantic Ocean and the Mediterranean Sea. The landlocked sea would have begun to shrink as its waters evaporated under the strong Mediterranean sun. With the increasing salinity of its waters, all marine lives would have died out, except for some dwarf species of clams and snails tolerant of very saline conditions. The inland sea would have eventually changed into a big salt lake, like a Dead Sea a hundred times magnified. The brine would have become dense enough to precipitate gypsum, but the evaporation would have continued. Finally, the Mediterranean bottom would have been laid bare. The nearby submarine volcano would have turned into a volcanic mountain. Oceanic oozes and gypsum deposits on its flanks would have become lithified. Streams draining such a terrain could have laid down a deposit of gravels such as the one we had found. Finally, seawater would have broken the dam at Gibraltar and flooded the Mediterranean basin. Where there had been a salt desert there would have been again a deep blue sea.

It seemed preposterous to spin such a tale on the flimsy bit of evidence we had. And Bill Ryan, for one, was extremely skeptical of my unorthodox idea. Ryan had worked with a "continuous seismic profiler," or CSP, which was a super echo sounder: besides recording sound echos bounced back directly from the sea floor, this instrument could send and pick up signals of acoustic waves that were able to penetrate the bottom and reflect off hard layers several kilometers below. The instrument had been developed in the late 1950s, and in 1961 Ryan sailed with his mentor, Brackett Hersey, on the American research vessel *Chain* from the Woods Hole Oceanographic Institution to explore the Mediterranean with the newly developed CSP. They soon discovered an acoustic reflector 100 to 200 meters beneath the Mediterranean sea floor. They had no idea what it could be or why it should be there, but for the sake of easy reference they named this mysterious layer the M-layer, and its top, the M-reflector. American and French scientists continued the CSP surveys of the Mediterranean during the next ten years, and wherever they sailed they could identify on their records the ubiquitous M-reflector. Furthermore, the geometry of this reflecting surface closely simulated the topography of the bottom of this inland sea (fig. 3); the sediments under the reflector covered the basement of the Mediterranean like a thick blanket of snow on a mountain plateau. Obviously the M-layer was deposited when the deep basin of the Mediterranean Sea had already been created and had almost the same bathymetry as it does today. Ryan and other geophysicists were

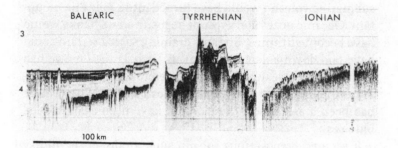

BALEARIC   TYRRHENIAN   IONIAN

100 km

3. Continuous seismic profiles of the M-reflector and the sea floor in the Balearic, Tyrrhenian, and Ionian basins in the Mediterranean; obtained by the research vessel *Robert Conrad*. Verticle scale is in seconds (two-way travel time of acoustic waves), and vertical exaggeration is about 5 to 1.

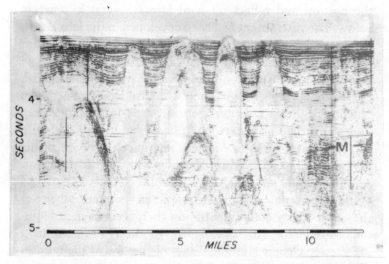

4. Continuous seismic profile of a ten-mile-wide section of the Balearic abyssal plain in the western Mediterranean. Some of the salt domes protrude as knolls above the sea floor; others are completely buried. This diagram is a copy of a record obtained by the French vessel *Calypso*. Courtesy of Olivier Leenhardt.

therefore convinced that the sediments making up the M-layer, whatever they were, had to be pelagic sediments, or fine sediments falling onto an uneven deep-sea floor like snowfall.

The seismic profiling surveys had also discovered that the Mediterranean sea floor is underlain in some parts by an array of pillarlike structures, each a few kilometers in diameter and hundreds or even thousands of meters tall, protruding into layered sediments (fig. 4). Geophysicists were familiar with this type of structure; they looked very much like salt domes, which are common along the United States Gulf Coast. Salt domes are formed when deeply buried rock salt forces its way upward into overlying formations. One would expect to find salt in coastal sediments, for evaporites had long been thought to have precipitated in coastal salinas or lagoons. One would not expect, however, to discover salt domes under the abyssal plains of the Mediterranean Sea. Some geologists, particularly those from the French school, thought that the salt must have belonged to a very old formation, some 200 million years of age, being mined on the continent of Europe. To find a salt deposit under the Mediterranean was thus considered evidence that the seabed was at one time a part of the continent but had sunk into the abyss like the proverbial "Lost Continent of Atlantis."

The discovery of gypsum-bearing gravels led us to a chain of deductions. We hit the gravels just at the top of the M-layer and thus thought perhaps the M-layer was a young evaporite formation deposited during the same epoch of geologic history as the salt in the salt dome. On this point Ryan and I agreed, but our different backgrounds led us to a crossroads: whereas my training led me to postulate that desiccation of the Mediterranean caused the salt and the gypsum to precipitate, Ryan's experiences in following the seismic record of the M-reflector convinced him that the evaporites had been laid down in a deep basin filled with dense brines.

Our low-keyed conversations soon turned into a kind of debate: Ryan convinced me that the evaporites must have been deposited in a deep basin, but there was no compelling evidence to suggest that the basin was full of salt-precipitating brines. He came back with a reprint from an article by a theoretical geochemist, Bob Schmalz of Pennsylvania State University. "Brine pockets" precipitating calcium sulphate had been discovered in deep depressions beneath the Red Sea. That appeared logical to a theorist, because brines should sink, being denser than normal seawater. Carrying the interpretation one step further, Schmalz reasoned that the Mediterranean Sea would become a deep "brine pool" if the evaporated seawater could be prevented from returning to the Atlantic Ocean. The highly saline water of the Mediterranean now flows out of the 400-meter-deep Strait of Gibraltar as bottom currents, to be replaced by an inflow of fresh seawater from the Atlantic. If the strait had been shallower once, so went the argument by Ryan quoting Schmalz, the Mediterranean could have been a deep briny sea.

I understood Ryan's logic. Evaporites could have been laid down on a deep-sea floor. But the gravels suggested that the Mediterranean evaporites were deposited on desert playas or in salt lakes, not in deep water. Admittedly the evidence was somewhat weak; we had not yet even hit an actual evaporite deposit, but only some erosional debris. We should wait until we got to the next drill site.

We did not run into the evaporite at our next site, but a pack of troubles. A few days later, however, we hit the jackpot in Hole 124. On the morning of August 28, the *Challenger* was drilling south of the Balearic Islands in almost 3,000 meters of water (fig. 2). Ryan and I had again stayed up into the early hours of the morning, when the drill pipe apparently hit the hard M-layer. The drilling rate dropped from several meters per minute to a meter per hour. Impatient with the slow progress, we went to bed just before dawn.

We were not to rest long. Soon we were awakened by John Fiske, a marine technician, who came to report: "We found the pillar of Atlantis!" We dressed quickly and rushed to the ship's laboratory to see the new find. Lying on the long worktable was a beautiful core, which did indeed resemble a miniature marble column (fig. 5). That was the evidence I needed.

Sedimentologists are students of sediments; they describe and analyze sediments and sedimentary rocks. They would cut a chip off a piece of carbonate rock, grind the chip into a transparent-thin slice, and examine this under a microscope. They would crush a shale, pulverize it, and bombard the powder with X-rays to determine its mineral composition. They would pound on a sandstone and shake it until the sand grains became loose enough to run through a series of sieves to analyze its size and sorting. They would dissolve an evaporite (a chemically precipitated rock) and process it through a mass spectrometer to determine isotopic ratios of various chemical elements. Their purpose is to learn more about the origin of a sediment. Is it a beach deposit, a lime mud laid down on a tidal flat, or an oceanic ooze?

In some instances one does not have to go through complicated procedures nor use sophisticated instruments; one can immediately tell the genesis of a rock by the way it looks. Techniques of comparative sedimentology were developed shortly after the Second World War, and the financial backing by the oil industry contributed considerably to their success. Teams were sent out to study Recent sediments in various environments: river sediments on coastal plains, deltaic sediments at the mouths of major streams, marine sediments on open shelves, oceanic sediments on abyssal plains, and so on. Distinguishing features were defined, then described as "sedimentary structures," and those

5. The "pillar of Atlantis"; the first core of evaporite deposits recovered from Site 124.

structures serve to characterize suites of sediments deposited at various places. When a core of an ancient sedimentary formation is obtained from a borehole or an oil well, one can now compare its sedimentary structures with a known standard, in much the same way an art historian identifies a purported Rembrandt by comparing its composition, coloring, shading, and brush strokes with known Rembrandts. Sometimes the comparison is purely empirical. Other times there are good theoretical reasons why a sediment should look the way it does.

Our "pillar of Atlantis," for example, consists of anhydrite and stromatolite. This type of sediment has been found only on arid coastal flats. Prior to the *Challenger* expedition, my associates and I at the Swiss Federal Institute of Technology, supported by a research grant from the American Petroleum Institute, had studied the sabkha sediments of the Arabian Gulf. We dug scores of trenches on the sabkhas of Abu Dhabi and found anhydrite, a calcium sulphate salt, only in those places where the saline ground water was sufficiently close to the surface to be heated to temperatures exceeding 30 degrees Celsius. Where the water table was deeper and the water cooler, gypsum, or hydrated calcium sulfate ($CaSO_4 \cdot 2H_2O$), would be precipitated out in place of anhydrite. This finding is in accordance with chemical studies in the laboratory, which reveal that the transition temperature for calcium sulfate precipitated from saline ground waters should be above 30 degrees Celsius, or almost 90 degrees Fahrenheit. We thus have good reason to believe that anhydrite is not likely to be found in any environment other than hot and arid sabkhas, because surface temperatures and ground water chemistry elsewhere rarely permit anhydrite precipitation. We are almost certain that anhydrite could not be settled out of a deep sea. Even the Dead Sea is too deep a body of water to be heated hot enough to precipitate anhydrite; on the bottom of this salt lake only gypsum crystals are found.

The anhydrite found under sabkhas was precipitated by ground waters like concretions in arid soils. Fine-grained anhydrite would accrete and grow together as nodules underground, replacing preexisting carbonate sediments (fig. 6, *lower left*). The nodules might range up to several centimeters in length. As the replacement proceeded toward completion, anhydrite nodules would coalesce to form a layer in which only wisps of preexisting carbonates could be discerned. The dark wisps of carbonate in a white background of anhydrite look like the wire mesh used by farmers to make chicken-wire fences (fig. 6, *lower right*). Thus petroleum geologists who first encountered such anhydrite in their study of borehole cores dubbed the rock type "chicken-wire anhydrite." We really do not know why anhydrite grows in this particular form. We can only rely on the repeated observations by sedimentologists during the last few decades that this variety of anhydrite is typical of Recent and ancient sabkha sediments. Until we find evidence to the contrary, we feel content to consider the chicken-wire anhydrite a signature of sabkhas.

Stromatolite is another distinct sedimentary structure (figs. 6, *upper right*, and 7, *upper right*). It had been considered a fossil or an inorganic structure of chemical precipitation until the 1930s when a British sedimentologist, Maurice Black, waded across the tidal flat of the Bahamas and found a dense growth of blue-green algae forming a thin mat on the flat shores (fig. 8). After a severe storm the mat would be buried under a thin cover of sediments, but the algal growth would persist and a new mat would be constructed. This alternation would ultimately result in the laminated sediment called stromatolite, which means literally "flat stone." Since the very existence of algae depends on photosynthesis, the presence of a stromatolite structure is con-

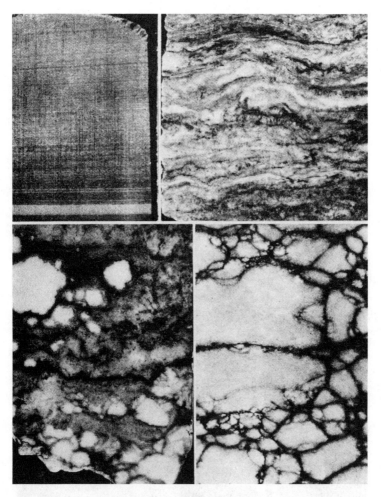

6. Mediterranean evaporites: (*upper left*) laminated marl; (*upper right*) stromatolite; (*lower left*) nodular anhydrite; and (*lower right*) chicken-wire anhydrite.

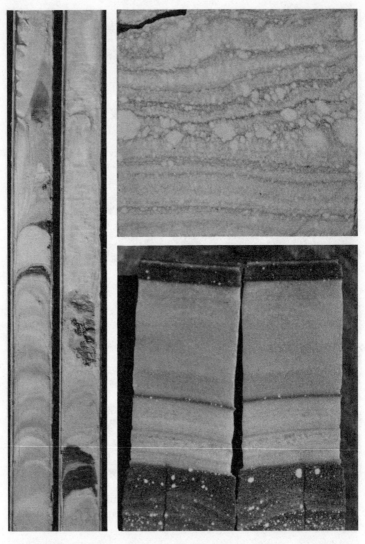

7. Sedimentary cores from the Mediterranean: (*left*) marine sediments made up of the dead bodies of very tiny animals (foraminifera) and even smaller plants (nannoplankton); (*upper right*) anhydritic dolomite, stromatolitic; and (*lower right*) salt core from the Balearic abyssal plain.

8. Blue-green algae cover the intertidal zone on the shore of an island in the Bahamas like a mat. The stromatolite shown in fig. 6 consists of fossilized algal mats.

sidered evidence of deposition in very shallow waters, commonly less than ten meters deep. In fact, repeated observations have confirmed that algal mats are a characteristic feature of intertidal environments. In the coastal areas between low and high tides, or the intertidal zone, of Abu Dhabi we found the current crop of lush growth in algal mats as well as old algal mats formed a few thousand years ago and now buried under the windblown sand of the coastal sabkhas. Transpiration of ground water led to precipitation of gypsum or anhydrite in these fossilized intertidal sediments (fig. 9). That August morning when I was called in to admire the "pillar of Atlantis," I saw the same phenomena of a stromatolite partially replaced by nodular anhydrite. What could be a better indication that these sediments were formed on the tidal flat of a desiccated Mediterranean?

The "pillar of Atlantis" was sampled from a layer of rock sandwiched between ocean oozes that contained abundant

9. Algal stromatolites and chicken-wire anhydrite from a trench in an arid coastal flat in Abu Dhabi.

fossil skeletons of foraminifera and nannoplankton. Maria
Cita, our shipboard paleontologist, was a professor from the
University of Milan and a specialist in planktonic foraminifera, tiny one-celled animals that range from a few tenths
to a few hundredths of a millimeter in size (fig. 10, *lower
left* and *lower right*). According to Cita, the plankton found
here once swam in the near-surface water of the oceans. After
they died, their calcium carbonate shells fell to the ocean
bottom and were buried and preserved as microfossils. The
nannoplankton were even smaller. They were unicellular
plants immersed in ocean waters, and they too secreted calcium carbonate skeletons in various shapes. Some of the
species (*Discoasters*) look like little snowflakes under a microscope, except that they are only a few thousandths of a
millimeter in size (fig. 10, *upper right*). The deep-sea floor
is a cemetery for billions upon billions of these tiny dead
plants; the skeletons of nannoplankton may constitute more
than ninety percent of the bulk of an oceanic ooze (fig. 11).
When these oozes are mixed with fine terrigenous particles
of clay, as they are in the modern Mediterranean, geologists
use the term "marl oozes," or simply "marls."

At the beginning of the nineteenth century an obscure
English land surveyor, William Smith, discovered that ancient strata could be characterized by the fossil shells one
finds in those strata. Several years later his countryman, Sir
Charles Lyell, found that Miocene and Pliocene sediments
in Italy contain different proportions of extant species. Micropaleontologists utilized the same principle to determine
the age of oceanic sediments, for tiny one-celled creatures
also came and went. One species after another would die
out only to be replaced by others, and the evolution of ocean
life left a record in the succession of different assemblages
of microfossils and nannofossils that characterize sediments
of different ages. Historians speak of dynasties when referring to the various eras in a country headed by different royal

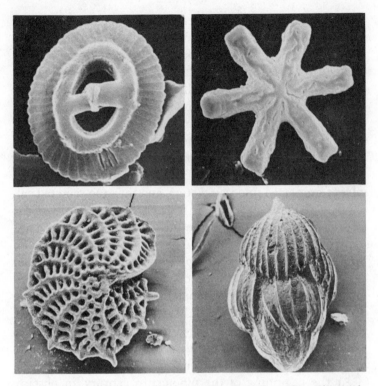

10. Some species of foraminifera and nannoplankton from the Mediterranean: (*upper left*) *Gephrocapsa* sp. (a nannofossil) *16,000* × ; (*upper right*) *Discoaster* sp. (a nannofossil) *4,000* × ; (*lower left*) *Uvigerina mediterranea* Hofker (a foraminifer) *75* × ; and (*lower right*) *Elphidium strigillatum* Fichtel/Moll (a foraminifer) *85* × .

families. Similarly, in the geological profession the concept of "fossil zones" is applied to define time intervals. For example, Cita found that the beginning of the Pliocene Epoch in the Mediterranean countries can be defined by the *Sphaeroidinellopsis acme* zone, or the time when the foraminifer genus *Sphaeroidinellopsis* reached the zenith of its development.

Paleontological dating gives us only the relative succession of sedimentary strata. Lyell used the terms "Miocene" and "Pliocene," or "more" and "less" recent, but he had no way of determining how recent. Physicists have since worked out a more precise method of dating, however, which entails measuring the amount of the parent and the daughter products of radioactive decay present in a rock. These results can be expressed in terms of millions of years. All one needs is a rock that contains radioactive elements such as uranium, thorium, potassium, rubidium, and so on. Most fossiliferous sediments do not contain the radioactive elements in abundance, but some are interbedded with volcanic ashes, or lava flows, which do. The latter can be dated and assigned ages in terms of millions of years, thus providing a calibration for the absolute age of fossil zones. The late Scottish geologist, Arthur Holmes, was a pioneer in establishing an absolute scale of geologic time. But the time scale has been much refined during the last decade, and we scientists in the Deep Sea Drilling Project were relying upon one provided by a colleague at Woods Hole, Bill Berggren, which places the end of Miocene at 5 million years before the present, and the end of Pliocene at 2 million years ago.

The "pillar of Atlantis" itself has not yielded any fossils, but the subjacent ooze contains planktonic foraminifera that lived only during an interval of geologic time called the Messinian. "Messinian" is a name used to designate a rock formation near the town of Messina in Sicily. About a century ago Professor Mayer-Eymar, one of my illustrious predecessors at Zurich, studied the fossils collected from a marl between some gypsum layers in this formation—which, incidentally, has been called *solfifera sicilienne*—and concluded that they were characteristic of an age just before the end of the Miocene Epoch. So he proposed a "Messinian stage" to designate this particular time interval. It was a relatively short stage, lasting perhaps less than 1 million

11. Oceanic sediments; taken with the aid of a scanning electron microscope, *4,000* ×. The ring-shaped fossils are skeletons of nannoplankton.

years. Our latest estimate would place the interval from about 6 to 5 million years ago.

Salt-bearing and gypsum-bearing formations like *solfifera* are common in other Mediterranean countries, like Spain, Algeria, Tunisia, Greece, Turkey, Cyprus, Israel, and so on. Many of these have also been designated Messinian.

The paleontological analyses by Cita thus enabled us to contradict the notion then prevailing that the Mediterranean salt bed was 200 million years old. The evaporites were deposited some 5 to 6 million years ago when the Mediterranean basin had already acquired a bathymetry similar to that of the present. I was already convinced that the deep basin was desiccated, but Ryan was more cautious. So each of us made a number of predictions. If the evaporites were deposits in local salinas and lagoons, they should be various from place to place, exhibiting none of the uniformity typical of deep-water sediments. If they were deposited on shores of shallow brine lakes, we should find remains of fossil plants, which required photosynthesis for their growth. If these brine lakes were situated in a continental environment, which may have been flooded occasionally, we should find non-marine or even freshwater fossils in the evaporites. If the evaporites were precipitating from progressive desiccation of the Mediterranean Sea, we should find a zonal arrangement with the least soluble salts, potash and magnesia, present in the deepest part of each Mediterranean depression where the last residue was being evaporated. If the Mediterranean Sea had been deep before its desiccation, we should find deep ocean sediment below the evaporites. If the deep Mediterranean bottom had been exposed to air, the base level of erosion of the Mediterranean drainage basins should be several thousand meters below sea level; the Mediterranean rivers would have cut gorges that extended from the present coast down to the abyssal plains. If the Mediterranean was once desiccated, there would have been easy migration of land animals from Africa to Europe and vice versa, and subsequent isolation of the immigrants stranded on Mediterranean islands when the Atlantic seawater reentered. If the Mediterranean sea floor was once a hot desert some 3,000 to 5,000 meters below sea level, there would have been large changes in the vegetational zones in Mediterranean lands in accordance to the changed climate. And so on.

These were the questions foremost in our minds during the remainder of the expedition, but our doubts were removed when we returned to Lisbon in October. The drill cores from the Mediterranean provided resoundingly positive answers to many of our predictions. It was an historic voyage, and this book is an attempt to retrace our steps: the planning, the execution, and the gradual revelation, leading to the formulation of a fantastic story. It is also an attempt to describe our life on *Glomar Challenger*, when we faced one crisis after another. Yet it was in the darkest moments of our despair that we cried "Eureka!"

1

GERM OF AN IDEA

───o───

SINCE the time of Isaac Newton, who determined the average density of the earth, it has been common knowledge that the earth's interior is made up of very dense matter and that there is an increase in density with depth. But in studying the transmission of seismic, or earthquake-generated waves, geophysicists have concluded that the increase in density is not gradual at every depth; there are discontinuities in the increases. One of these, discovered by the Yugoslavian geophysicist Mohorovicic, is about five kilometers below the ocean floor and a few tens of kilometers below the surface of the continents, where the density changes abruptly from 2.8 to 3.3 or 3.4 g/cm³. This discontinuity was named the "Mohorovicic discontinuity" to honor its discoverer but is better known in its abbbreviated form, "the Moho." The part of the earth above the Moho is called the "crust" and the part below is called the "mantle"—a 2,900-kilometer thick layer mantling the core of the earth.

In the early 1950s a group of prominent scientists in the United States hit upon the idea of drilling on the ocean bottom, where the Moho might be more easily reached, thereby gaining some insight into the nature of the Moho and the rocks in the mantle. This project of drilling through the Moho became known as "Project Mohole." Many of us did not think the idea practical. Nevertheless, with the blessing of Congressman Albert Thomas, chairman of the Appropriations Committee of the United States House of Representatives, millions of dollars were funded for this ambitious undertaking. A few holes were drilled, one reaching several hundred meters depth, but all were short of the goal. Academic and congressional opposition eventually killed the Mohole project shortly after the demise of Representative Thomas. But this abortive trial did develop a most important "dynamic positioning system," which could keep an unanchored drilling vessel at a given spot while the drill string bored its way into the ocean bottom (fig. 12).

The Mohole project was not a bad idea. It might even have been successful if the goal had justified a crash program or an immodest investment of funds and manpower. In the aftermath, however, when oceanographers got together and talked about drilling the ocean bottom again, they considered more modest objectives. The outcome of such discussions was the Deep Sea Drilling Project. The original proposal was a joint effort by scientists from four major oceanographic institutions in the United States: The Lamont-Doherty Geological Observatory of Columbia University, The Rosenstiel School of Marine and Atmospheric Sciences of the University of Miami, The Scripps Institution of Oceanography of the University of California, and the Woods Hole Oceanographic Institution. These four constituted the original JOIDES, the Joint Oceanographic Institutions Deep Earth Sampling program. Later the University of Washington became the fifth member. The proposal was eventually approved, and National Science Foundation funding was granted to the Deep Sea Drilling Project of Scripps

12 (on facing page). Dynamic positioning system of the Glomar Challenger. Maximum allowable drift is about 3% of the drilling depth (or 150 meters for 5,000 meters drilling depth). A normally functioning system could easily maintain the position of the vessel within a radius of 50 meters. Courtesy of DSDP.

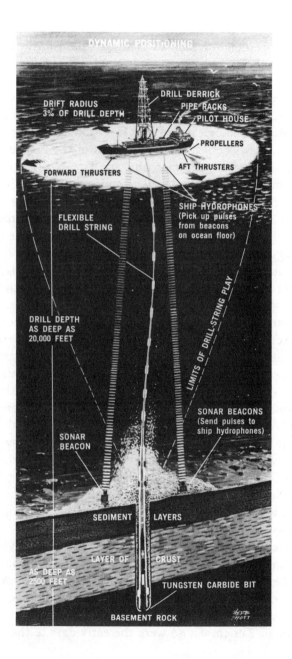

DRIFT RADIUS
3% OF DRILL DEPTH

DRILL DERRICK

PIPE RACKS

PILOT HOUSE

PROPELLERS

FORWARD THRUSTERS

AFT THRUSTERS

FLEXIBLE
DRILL STRING

SHIP HYDROPHONES
(Pick up pulses
from beacons
on ocean floor)

DRILL DEPTH
AS DEEP AS
20,000 FEET

LIMITS OF DRILL-STRING PLAY

SONAR BEACONS
(Send pulses to
ship hydrophones)

SONAR
BEACON

SEDIMENT LAYERS

LAYER OF CRUST

AS DEEP AS
2500 FEET

TUNGSTEN CARBIDE BIT

BASEMENT ROCK

as the operator for JOIDES. The inaugural drilling cruise
sailed from Orange, Texas, on July 20, 1968.

At about the same time I was on my way from Zurich to
Prague. In August 1968 the quadrennial sessions of the International Geological Congress met at Prague. I was among
those who witnessed the last blooms of the Prague spring
and saw the Russians marching in. The experience left a
deep impression on all of us who were there. In the September aftermath some Czech colleagues sought temporary refuge in Zurich. Since Switzerland provided little in the way
of employment opportunities for them, we all wrote to our
colleagues overseas. I wrote Jerry Winterer at the Scripps
Institution, inquiring if the newly started Deep Sea Drilling
Project would need someone for temporary employment.
The response was immediate and positive. But the offer came
after our colleagues had received and accepted another offer.
Appreciating their circumstances, I wrote a long letter to
Winterer, thanking him for his trouble and apologizing for
my Czech friends' not being able to accept his generous offer.
At the end of the letter, I added: "Of course, if I should have
left you in a pinch, and if you would need anyone on short
notice, don't hesitate to call upon my service."
    That was in October 1968. At the beginning of November
I received an unexpected telegram from Scripps, which asked
me if I could and would participate in the third leg of the
Deep Sea Drilling Project to the South Atlantic. The expedition was to leave in a few weeks. I was caught completely
by surprise.
    Sometimes I wonder if Winterer extended this invitation
with the intention of teaching me humility. We had had a
very heated scientific argument during a farewell party for
me in early 1967, when I was to leave the United States to
accept a position with the Swiss Federal Institute of Technology in Zurich. Those were the early days of the "earth

science revolution," which was to change radically how we regard the history of the earth. The revolution was a direct consequence of the postwar exploration of the oceans, roughly three quarters of the earth's surface. The new findings could not be explained by old dogmas, which had been established largely on the basis of our knowledge of the continents.

A routine survey being carried out on almost every oceanographic cruise is to measure the intensity of magnetization on the sea floor with a shipborne magnetometer. Unusually high magnetization is referred to as a magnetic anomaly. Magnetic anomalies on land are as a rule irregular in distribution and are related to high local concentrations of magnetic minerals in a rock. Magnetic surveys have thus been a common tool used in the search for iron ores. Magnetic surveys of the ocean floors during the 1950s came up with two surprises: magnetic anomalies on the sea floor are not only more intense than those commonly found on land but they have also a linear arrangement, forming so-called magnetic stripes.

When a young graduate student at Cambridge University, Fred Vine, joined an oceanographic cruise to survey the Indian Ocean in 1962, magnetic stripes had already been delineated in many places. Vine was not satisfied with just confirming this pattern by recording his observations; he wanted to know why the anomalies were arranged in this way. The stripes did not seem to correlate with local submarine topography. Rather, their orientation was parallel to the trend of features that have been called mid-ocean ridges (fig. 13). Furthermore, he noted that the anomalies were symmetrical with respect to the axis of the ridges (fig. 14). As a student of geophysics, Vine was familiar with two popular hypotheses in earth sciences. A new controversial hypothesis had suggested that the earth's magnetic poles have been repeatedly reversed during the last few hundred million years. The other, an old and in the United States almost discredited

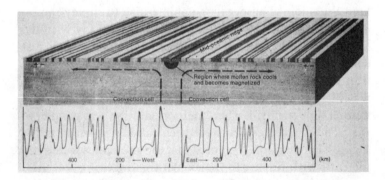

13. Diagram of the Mid-Atlantic Ridge. The ridge has a height of 2,500 meters below sea level, whereas its flank is about 5,000 meters deeper. After a diagram by Marie Tharpe and Bruce Heezen.

14. Diagram of magnetic stripes on the sea floor; dark stripes and high positive anomalies (peaks in curve below the diagram) represent magnetization at times when the positive magnetic pole and the north geographical pole approximately coincided. Courtesy of DSDP.

hypothesis, contended that the ocean was born when continents drifted apart from each other. "I was only trying to combine two well-known ideas," Vine was to tell me years later. With the genesis of the oceans, continents moved apart to make room for lavas coming out of a mid-ocean ridge; the cooling lavas that formed a stripe of new sea floor there were magnetized according to the prevailing magnetic field. After some time the magnetic poles reversed themselves and the lavas coming out of the ridge formed a new strip of sea floor with a reversed magnetic polarity. The separation of continents and the spreading of sea floors during repeated polarity reversals seemed to provide an explanation for the magnetic stripes.

Vine and his thesis adviser, Drummond Mathews, published this idea in 1963, and Vine gave an exposition of his new hypothesis later in 1966 during the San Francisco meeting of the Geological Society of America. The story soon became the talk of the cocktail circuit.

I had been skeptical of Vine's elegant exposition and voiced my conservative opinions during the farewell party mentioned earlier. Jerry Winterer, on the other hand, prophesied that the new creed would revolutionize our thinking. After a heated debate he swore that he would see to it that I ate my words. And he did.

Correlating the width of the magnetic stripes with the duration of successive reversals of magnetic poles, the sea floor spreading hypothesis should provide a means of determining the age of the ocean floor: the farther away from the ridge axis, the older the ocean crust would be, and the ratio would measure the rate of sea floor spreading. The Leg 3 DSDP expedition to the South Atlantic was planned to test the hypothesis. By drilling, sampling, and dating the ocean crust, we should find out if the ocean floor at a number of chosen sites was indeed as old as the hypothesis predicted.

Working on *Glomar Challenger* during the Leg 3 drilling,

I witnessed the most amazing confirmation of this concept of sea floor spreading. We bored ten holes, and the age of the sea floor at every site was almost exactly that predicted by the hypothesis (fig. 15). It is always hard for me to accept other people's brilliant ideas and admit my own errors in judgment, but faced with ironclad proof, I had no choice but to join the "revolutionaries."

One of the chief scientists on the South Atlantic expedition was Art Maxwell, then the director of research at the Woods Hole Oceanographic Institution. He had been an active member on several JOIDES panels, which planned and advised the deep-sea drilling program; the DSDP simply administered the grant from the National Science Foundation as JOIDES directed. The first grant was to end in February 1970. During the Leg 3 cruise in the last days of 1968, we shipboard scientists chatted often with Maxwell about the future of the project. He was happy with the success of the first three legs and was confident that the project would be continued into the 1970s.

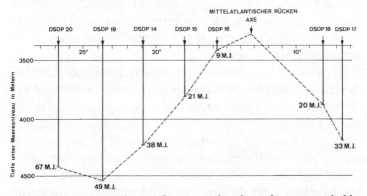

15. Confirmation of the sea floor spreading hypothesis provided by results from the Leg 3 deep-sea drilling cruise to the South Atlantic in 1968.

I, as a land geologist, had a particular interest in seeing some boreholes drilled in the Mediterranean. Geologists had long believed that the sedimentary rocks that form mountain chains were deposited in geosynclines or small oceans. The sediments of the Himalayas and the Alps, for example, were thought to have been deposits in a geosyncline that was called the "Tethys," the daughter of Gaia (Earth) in Greek mythology. The present Mediterranean, then, would be a small remnant of the Tethys geosyncline, after the rest of its sediments had been compressed and squeezed up to form mountains. With the progress of the recent revolution in earth sciences, however, the old thinking on geosynclinal development was being replaced by a new theory of plate tectonics. This new theory was the work of an international group of young scientists, including, among others, Jason Morgan of Princeton, Dan McKenzie and Bob Parker of Cambridge, and Xavier Le Pichon, then at Lamont, under the influence of their mentors, Harry Hess, Teddy Bullard, and J. Tuzo Wilson. The old dogma had contended that continents and oceans remained fixed and stable in their positions throughout the history of the earth, while the earth's crust in more mobile belts bent down to form geosynclines and buckled to form mountains. The new school believed that the earth's crust was composed of a number of rigid "plates," each moving in relation to others. New ocean crust was added at the boundaries of plates moving away from each other, and mountains were formed where moving plates came together and collided (fig. 16). Thus the mountain building process had nothing to do with geosynclines. The Tethys and the Mediterranean in the framework of plate tectonics were not geosynclines but the boundary between the European and the African plates, moving at times away from and at times toward each other.

Since the Mediterranean Sea had a place in both the old

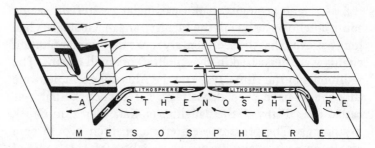

16. Schematic diagram of plate tectonics theory; originally prepared by Brian Isacks, Jack Oliver, and Lynn Sykes of the Lamont-Doherty Geological Observatory. Courtesy of the American Geophysical Union.

and the new theories, it seemed to me that a few boreholes in this inland sea might go a long way toward resolving the conflicts and misunderstandings. The Mediterranean, however, is underlain in most places by a thick sedimentary blanket, and our drilling technique was not sufficiently advanced to insure penetration to the basement of a basin, where most of our answers should lie. A drilling cruise to the Mediterranean was thus opposed by several prominent geophysicists on the JOIDES panels. Fortunately, during the long months at sea, Art Maxwell lent me his ear. He was to be one of our strongest supporters.

In April 1969 I went to Scripps for a conference in connection with the South Atlantic venture. There I mentioned to Jerry Winterer that I thought we ought to include an expedition to the Mediterranean if DSDP should be extended. He was sympathetic and advised me to contact Bill Riedel, the Scripps representative on the JOIDES planning committee. So I wrote Riedel after I returned to Zurich and sent him a proposal.

In June 1969 I met Art Maxwell again at Woods Hole. We spent two days together working on our South Atlantic expedition report. In spare moments, however, I did my best to lobby for the Mediterranean drilling expedition, for I had learned that Maxwell had become the chairman of the JOIDES planning committee. By then the extension of the DSDP program had been confirmed, and I was given the impression that a voyage to the Mediterranean was indeed likely. In fact, an advisory panel was about to be set up to plan for Mediterranean drill sites. There the matter stood.

2

FRIENDS OF JOIDES

——o——

IN OCTOBER 1969 I received a letter from Art Maxwell on behalf of the JOIDES executive committee, nominating Brackett Hersey, Bill Ryan, Bob Hurley, and me as members of the JOIDES Mediterranean advisory panel. At about the same time I received a phone call from some industry people in France, asking for an appointment; they had sent drilling proposals to Lamont and were told to contact me in Zurich.

The Deep Sea Drilling Project was an American program. The National Science Foundation supported it for eighteen months with an original grant of $12.6 million, and the first three-year extension was budgeted at $35 million. The project welcomed the participation of individual European scientists (then in 1975 formally invited European institutions to join it in forming the International Phase of Ocean Drilling [IPOD]), but the DSDP cruises during the first eighteen months had been scheduled to drill in the Atlantic and the Pacific, somewhat remote from Europe. Now that the DSDP was planning an expedition to the Mediterranean, it would be discourteous to European colleagues if Americans were to poke holes in their backyard without their active participation. For this reason, I told Maxwell that I would try to use my geographic situation to bring European friends of JOIDES together, Zurich being located in the heart of Europe.

In early November we had the first meeting of the friends of JOIDES. Several oceanographic institutions in Europe sent representatives to Zurich, and industry people joined us there. We had little knowledge of the capabilities of the drilling vessel in those early days. People in industry thought she might be used to secure information for oil exploration and wanted us to drill on top of salt domes. Thus I soon found myself under heavy pressure from petroleum interests. At this juncture I was pleased to receive the support of Xavier Le Pichon, a young star of French oceanography who was then the head of their oceanographical research center at Brest. He spoke out and rescued me from the siege by his countrymen, pointing out that the purpose of deep-sea drilling was to investigate the origin of ocean basins, not to get lost in trivialities; the proposal by industry, with its emphasis on salt structures, did not conform to the objectives of the Deep Sea Drilling Project. With this clear definition of our aims, we settled down to our task, and by the end of the day, we had singled out several areas of interest, where we might drill to the bottom of sedimentary deposits and reach basement.

On the other side of the Atlantic another member of the Mediterranean advisory panel, Bill Ryan, was most active. Ryan, with his four previous Mediterranean expeditions, was one of the oceanographers most knowledgeable about the region, even though he was then not yet thirty years old and had just completed his Ph.D. thesis. He wrote me and invited me to meet him at Lamont to coordinate our efforts.

Although Ryan and I had corresponded, our paths had never crossed. I decided to make the trip in December. Arriving in New York, I telephoned my friend Tsuni Saito, with whom I had worked on the *Challenger* expedition to the South Atlantic. He picked me up and took me to his home for a Japanese dinner, telling me on the way home that he had also invited Ryan to dinner. I was quite pleased, for I was eager to meet this person with whom I was soon to work so closely. Saito and I had barely arrived before Ryan showed up. The surprise was mutual. He expected a gray-bearded European professor, and I expected a go-go type who would make the Junior Chamber of Commerce's list of Ten Young Men Most Likely to Succeed. Yet Ryan was soft-spoken and thoughtful. To my embarrassment, Ryan and I became oblivious of our gracious host and hostess and of the delicacies that had been so painstakingly prepared. We became engrossed in a marathon conversation that lasted almost twenty-four hours, interrupted only by a short night of rest. It was a productive session. I told Ryan the results of our Zurich meeting, and he brought out numerous profiles and diagrams and involved me in the technical details of pinpointing drill sites.

Meanwhile I began to learn a few things about the Mediterranean. It seems that the Mediterranean basin consists of two distinct entities. The east is characterized by a submarine mountain, commonly known as the Mediterranean Ridge, which rises some 2,000 meters above the abyssal plain (fig. 1). North of this east-west trending ridge is the Hellenic Trench, where the Mediterrranean plunges to its greatest depth of more than 5,000 meters. Still farther north is an arc of peninsula (Peloponnesus) and islands (Ionian Islands, Crete, and Rhodes), a feature that has been called an "island arc." Island arcs are regions of high earthquake activity. Behind the arc are volcanoes, and the Santorini in the Aegean is known to have caused repeated damage to ancient civilizations. Advocates of the plate tectonic theory had developed a neat explanation for island arcs and ocean trenches. An arc was supposed to be the fringe of an "active plate margin," where moving plates are coming together. The boundary between the European and African plates, for example, was supposed to lie at the foot of the steep wall of the Hellenic Trench (fig. 17). As Africa marches northward, the theory maintained, the trench floor is being tugged under the arc in a process known as subduction. The downward movement of the African plate leads to earthquakes, and the subducted plate, melted partially by the subterranean heat, furnishes the lava for active volcanism. The theorists told us that this was a place where mountains were presently being built. The information gathered by deep-sea drilling in the Mediterranean was to clarify our understanding of this important process.

The eastern Mediterranean was apparently being compressed and seemed to be what was left of a much larger former ocean, the Tethys, as Africa had been moving toward Europe. The western Mediterranean, on the other hand, was apparently being extended. Half a century ago a well-known Swiss geologist, Emil Argand, suggested the fascinating idea that Italy had been split away from the coast of Spain and had rotated more than sixty degrees counterclockwise until she met the Balkan block. The collision gave rise to the Apennine mountains, and the hole left behind by the drifting "boot" represents the western Mediterranean. According to this imaginative master, Corsica and Sardinia went only half

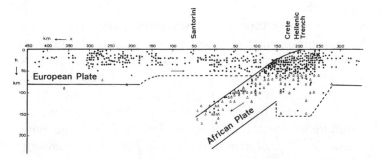

17. Cross section of the Aegean region, showing the earthquakes associated with the subduction of the lithospheric plate. Courtesy of the American Geophysical Union.

way, so they now serve as a partition between the Balearic
and the Tyrrhenian basins (fig. 1). Since the movement was
believed to have started some 25 to 30 million years ago,
the western Mediterranean was presumed to be a young geologic feature, formed mainly during the Miocene. We hoped
to test this theory by planning a few boreholes to ascertain
the age and structure of the western basins.

Ryan came to Zurich in February 1970 for the second
meeting of the European friends of JOIDES, a group that
included among its ranks the directors of several European
oceanographical institutions. On Friday the twentieth he
presented the first draft of his drilling proposal to this august
body. He reminded me on this occasion of the young Eisen-
hower, a light colonel who was to become the Supreme Com-
mander of the Allied Forces. With his youth and his apparently lethargic monotone, however, Ryan caused some raised
eyebrows. As we were to learn later, more than one person
was concerned that the planning of this important expedition was being left in the hands of a student (Ryan) and an
amateur (me). Nevertheless, this was an American undertaking, financed and organized by American institutions,
and Ryan and I were deemed the best available Americans
for the job.

On the whole, this meeting was not a failure. The French,
for example, opened all their files of unpublished data to us;
they further promised to help in the matter of site surveys.
As events were to prove, seven out of the fifteen sites that
we eventually drilled were pinpointed on the basis of an
excellent survey by Le Pichon's group at Brest. We were also
fortunate to have the chief scientist of the French surveying
cruise, Guy Pautot, join us on the *Challenger* as a sedimentologist. His participation contributed considerably to our
eventual success.

3

AUGUST 13, 1970, LISBON

———o———

THE SPRING of 1970 was a hectic time for me. Having been elected chairman of the Advisory Committee on Geophysics Appointment at the Swiss Federal Institute of Technology, I was to play host to leading candidates who visited Zurich as distinguished lecturers. Much of my energy was thus devoted to scheduling lectures, locating auditoriums, making travel arrangements, phoning, writing letters, changing plans, and so on. Adding to my woes, I had consented to contribute to three invited papers and was trying desperately to meet deadlines. Finally, I had to keep the research of our laboratory going, not to mention teaching a graduate and an undergraduate course that semester. Amidst all of this turmoil, I received a telegram from Scripps in late April, inviting me to be co-chief scientist on Leg 13 of the *Glomar Challenger* expedition to the Mediterrranean. Bill Ryan was to be the other co-chief scientist.

I attended to my everyday commitments until the time finally came for me to leave for the Mediterranean. Then on August 10 my whole family drove me to Kloten Airport and waved good-bye as I stepped on the plane to Lisbon, where I would meet the *Challenger*. Two paleontologists also destined for Leg 13, Wolf Maync from Bern and Herb Stradner from Vienna, took the same flight. We were greeted by sunny skies in Lisbon and were among the first to check in at the Florida Hotel.

18. The *Glomar Challenger*. Courtesy of DSDP.

Lisbon is a harbor town, reminding one of Naples or Genoa. The city was almost completely destroyed by a tsunami flood generated by the 1755 earthquake. Almost all of the present buildings in the city have been erected since. Yet people are short on memory and often ignorant of the origins of natural catastrophes. The Lisbon earthquake was caused by movement along a large fault zone, the Azores-Gibraltar fracture. (In fact we were to drill our first hole near this fracture to explore its geologic history.) Movements on active faults are recurrent, and history could repeat itself. Yet the people rebuilt the city on the flat bottom of the former creek bed that was flooded during the last catastrophe. The skyscrapers on both sides of the main thoroughfare would surely be wiped out if an earthquake-generated tidal wave should revisit the city. I tend to believe one of the geophysicists in the local university, who estimated that the death toll might rise past the million mark should the 1755 earthquake recur today. Yet the humming humanity through which I threaded my way was apparently oblivious to the calamity that could befall it at any moment.

When I returned to the Florida, I saw a large group gathered in front of the hotel. Wolf Maync was there, talking to Maria Cita and her assistant, Isabella Premoli-Silva. Others included the ship's crew, administrators from the Deep Sea Drilling Project, and Ken Brunot, the project manager at the time. They were all discussing the plight of Paulin Dumitrica, a micropaleontologist from Romania. Apparently, Terry Edgar, the project's chief scientist then, had received news from La Jolla that our Romanian colleague had had trouble securing a visa to enter Portugal.

While I waited for my breakfast the next morning, a man in a dark suit and somber mood approached me and introduced himself. He was Vladimir Nesteroff, a sedimentologist from the Sorbonne. We shook hands, then had breakfast together before the others came. Herb Stradner had gotten up early and was already on the dock when the *Challenger* arrived. The rest of us drove down to the harbor after eating. We could see the ship's sixty-meter drilling tower even from a distance (fig. 18).

The *Glomar Challenger* is an 11,000-ton drilling vessel, built especially by the Global Marine Company for the use by the Deep Sea Drilling Project. At the time it was the only vessel of its kind capable of drilling a 1,000-meter hole in 6,000 meters of water. The drilling tower has a load capacity of half a million kilograms, or the weight of more than 7,000 meters of drill pipe. The tower rises above a hole in midship, the "moon pool," through which the assemblage of drill pipes, or the drill string can be lowered down to the sea bottom. More than ten kilometers of drill pipes are stored on the racks forward of the tower (fig. 19), and the pipes are fed automatically to the drilling rig. A tiny hut was built on the starboard side of the tower to house the instrument panels and equipment (fig. 20). Unbeknownst to me then, I was to spend most of my waking hours during the next two months in this driller's shack.

The working and living quarters are housed in the stern of the vessel. Specially designed laboratories for micropaleontology ("the paleo lab") and for sedimentology ("the core lab") occupy successive decks above the engine room, while a refrigerated "core storage" is on the bottom deck. Farther astern, under the bridge, are dining room and ship offices. Toward the very back are bedrooms and lounges for the scientists and crews. The vessel can house ten to twelve scientists, about an equal number of technicians, plus a dozen members of the drilling crew, and she is served by more than thirty sailors and officers. Enough provision can be stored so that she can be at sea for two months without having to return to port.

After a year and a half, I was happy to be on board again, and I was greeted by many old acquaintances from the crew

19. From the top of the *Challenger*'s drilling tower, looking forward at the racks of drill pipe. Courtesy of DSDP.

20. The drilling crew rigs hydraulically operated power sub (center), preparing to commence drilling operations. At right is the driller's shack, with corrugated sheet metal sides, and to the left is the specially designed core line reel used to retrieve the core barrel. Courtesy of DSDP.

who sailed with Leg 3 across the South Atlantic. Stradner immediately went down to the paleo lab to check the equipment there. I went to the "science office," where the files were kept, being eager to find out what the last cruise had dug up. Later in the morning Bill Ryan arrived via "jumbo jet," a new toy then. We had much to talk about.

Ryan had visited La Jolla shortly before to secure approval

of our drill sites from the JOIDES safety panel. In 1970 JOIDES consisted of scientists from five American oceanographic institutions (the original four plus Washington). It was governed by an executive committee, and technical decisions were made by a planning committee; both of these committees had a representative from each of the participating institutions, with an ex officio member from the Deep Sea Drilling Project. Reporting to the planning committee were a number of panels. Three major panels at the time were the Pacific, the Atlantic, and the Indian Ocean panels, planning for drill sites in those three oceans respectively. The Mediterranean advisory subpanel was, for some strange reason, a subcommittee of the Indian Ocean panel, but we acted rather independently and had reported directly to the planning committee. In addition to the so-called geographic panels there were the "specialty panels" that dealt with specific problems. The safety panel was a newly constituted specialty panel whose purpose was to ascertain that DSDP scientific drilling would be "safe"—in other words that it would not encounter any oil or gas to cause possible pollution of the seas.

Apparently the political climate had changed drastically since the first meeting of the European friends of JOIDES. Students, conservationists, newspapermen, and politicians had all become ecology conscious. Meanwhile both the National Science Foundation and the Deep Sea Drilling Project had become acutely aware of the danger of an accidental encounter with hydrocarbons. Many of the Leg 10 drill sites in the Gulf of Mexico, for example, were cancelled because of the possible danger of pollution. Furthermore, in the new fervor for protecting the environment, some of those who had once favored drilling on salt domes had turned completely around; they now advocated the cancellation of the Mediterranean expedition because it seemed that large petroleum reserves might indeed exist in the Mediterranean.

Those colleagues whose interests lay in the Atlantic were proposing an alternative cruise from Lisbon to Monrovia. Fortunately the counsel of our supporters had prevailed at the last planning committee meeting, and the Mediterranean leg was approved on the condition that the proposed drill sites were not considered safety risks by the newly constituted safety panel. So, at the last minute Bill Ryan and Roy Anderson, who was to be our operations manager, met the safety panel at La Jolla. They went over the drill proposal site by site before the panel was satisfied. As Ryan and I now sat there in the mess hall of the *Challenger*, chatting, we were given a copy of the telegram:

etat k e brunot project mgr care d a knudsen
and co ltd cats do sodre 8-2 lisbon

leg approval granted substance and format as
previously again contingent upon following
recommended safety precautions formal approval
went to nierenberg

j d sides acting head national centers and
facilities operations

So we finally had our "visa" to the Mediterranean.

On August 12 the *Challenger*'s captain, Joseph Clarke, held an orientation meeting attended by all shipboard scientists, project personnel, the ship's officers, the drilling supervisor, and the operations manager. After Captain Clarke had briefed us on all essential operational information and cleared up the channels of communication, I finally raised the question of departure time. He replied that it was up to the co-chief scientists to decide. I was surprised that this decision was our responsibility. Eager to get to work with a minimum loss of time, I proposed a morning departure. No, that was not possible; they had to load some more supplies.

"How about early afternoon?"

"We have visitors who want to see the ship tomorrow evening. They would not leave until nine or ten o'clock."

"Do we have to have them?"

"Yes, the National Science Foundation has made the arrangements."

So it turned out that we could not leave before 2200 hours on the thirteenth.

The ship's log eventually gave a departure time of 0001 on the fourteenth. To this day I am not certain if the departure was delayed on account of the mariner's superstition. In fact, I myself was somewhat uneasily aware that our voyage was designated Leg 13 and that we were scheduled to sail on August 13; I had checked the calendar to reassure myself that it was at least not a Friday. That spring the incident of Apollo 13 had added fuel to the fire of numerologists. And once I was on board ship, everyone seemed to be warning me that it was unlucky to leave on the thirteenth. "Look what trouble Leg 10 had," somebody said. "They left on Friday the thirteenth and were barely on location before they had to return to Brownsville to repair the thrusters of the ship."

Others enumerated Leg 10's further misfortunes. The ship's captain was fired after that cruise, perhaps because he forgot to make a turn and went the wrong way while the chief scientists slept. The choice objectives could not be drilled because of some new rulings by the National Science Foundation. Stuck drill pipes, lost equipment, and so on. It was perhaps not unintentional that visitors had been scheduled to delay our departure until the fourteenth.

We had a meeting of the scientific staff in the afternoon of the twelfth. There were nine of us: Bill Ryan, Vladimir Nesteroff, Guy Pautot, Forese Wezel, Jenny Lort, Maria Cita, Herbert Stradner, Wolf Maync, and me (fig. 21). Only Paulin Dumitrica was missing. As the chief scientist of the Deep

21. The scientific staff of the DSDP Leg 13 expedition to the Mediterranean. Standing from left to right: Ryan, Pautot, Wezel, Stradner, Nesteroff, and the author. Sitting, from left to right: Maync, Lort, and Cita. Courtesy of DSDP.

Sea Drilling Project, Terry Edgar had had the task of assembling a scientific staff for the Leg 13 cruise; his nominations were subject to approval by the JOIDES planning committee, and by the National Science Foundation. Ryan and I were chosen shortly after we made our drilling proposal. As prospective co-chief scientists, we had had the privilege of making some suggestions for the rest of the staff. We nominated Pautot of the French Oceanographic Research Center at Brest, Wezel of the University of Bologna, and Lort of Cambridge University as sedimentologists; they each came from institutions currently active in the exploration of the Mediterranean. Nesteroff, of the Sorbonne, had applied directly to the Deep Sea Drilling Project. Bill Riedel and his JOIDES panel on biostratigraphy recommended our team of micropaleontologists: Cita of the University of Milan, Stradner of the Austrian Geological Survey, Maync, a self-employed consulting geologist, and Dumitrica of the Romanian Academy of Sciences. Terry Edgar thus had given us an international team of scientists from Austria, France, Great Britain, Italy, Romania, and Switzerland. The only participant from an American institution was Bill Ryan.

After we had defined our objectives, assigned duties, and cleared up routine details during this first staff meeting, Ryan proposed an addition to our agenda. He took out of his bulging briefcase a pencil-scribbled proposal made by our friend at Brest, Xavier Le Pichon, to drill on the Gorringe Bank, in the Atlantic southwest of Lisbon. I felt much irritated. The Deep Sea Drilling Project had already made six cruises to the Atlantic and had a couple more on the schedule. Why should we sacrifice *our* ship-time to punch one more hole in the Atlantic when all of us were eager to get east of Gibraltar to break new ground in the Mediterranean? Furthermore, I did not particularly appreciate the purpose of the proposal. So Ryan and I started our partnership on this discordant note. We were still studying the plans when the

VIP's showed up. Ryan took the American ambassador and his retinue for a tour of the ship, while I stayed in the captain's office talking with the wife of the naval attaché, who did not want to try her high heels on the ship's slippery stairways.

After our distinguished guests had left, Terry Edgar cornered me and asked me to join him and several others in trying to telephone Romania to inquire about Dumitrica. The call did not go through, however, so we all drove out to dinner. Once there, Ryan and Edgar tried again to persuade me to go along with them on the Gorringe Bank proposal. I was not convinced, but I began to give in for the sake of preserving a harmonious working relationship with Ryan. Besides, Le Pichon had done much for us and I hated to disappoint him too. On the way back to the ship, I finally blurted out:

"Okay, Bill, you win. We go to the Gorringe Bank. But we are sure to hit a hard bottom there and break the bottom-hole assembly."

Ryan replied, with a hint of melancholy:

"Don't wish me bad luck, Ken. There are enough people at Lamont who would be happy to say 'I told you so.'"

So the Gorringe Bank hole in the Atlantic was drilled. And it turned out to be one of the most interesting of our sites. Later when I thought of the episode, I reminded myself not to depend so stubbornly on my own intuitive judgment. Ryan, for his part, in the many talks he gave after the cruise, would claim that the hole was bought with 150 escudos, referring to the dinner bill they had paid for me because I had run out of local currency. An even better story was told during the Bay of Biscay Symposium in Paris in December 1970, when I was alleged to have gulped down so much wine in the restaurant in Lisbon that I did not wake up again until we were on our way to the Atlantic site.

We all went back to the Florida Hotel after dinner to make

one last try to phone Romania. The connection this time was made, but the person at the other end of the line could not be understood. After half an hour of total confusion, we gave up. But Bob Gilkey, the logistics officer from the DSDP, was determined to find Dumitrica. Just before Ryan and I stepped onto the gangplank, he asked us to estimate the timing of our passage through Gibraltar; perhaps he could deliver the missing Romanian scientist to us there.

It was soon midnight. We all leaned on the railing and waved good-bye to our friends as the *Challenger*, with her lateral thrusters in action, sidestepped from the dock of Lisbon.

4

CRISIS AND TRIAL

—o—

THE DERRICK of the drilling ship was forty-five meters above
the rig floor and almost sixty meters above the water line.
It was quite a sight to watch this tower slip under the Lisbon
Bridge. We were piloted slowly down the Tagus and were
finally in the open sea around two in the morning. I was
impatient, however, and the short trip to our first site seemed
a long wait. Suddenly I had the feeling the ship was not going
anywhere at all. I went to the bridge to check. And sure
enough, we were idle, waiting for the pilot to be picked up.
"Where is the pilot boat?"
"I don't know. She was behind us for awhile, but now
she's disappeared."
So we got off to an odd start. It turned out that the pilot
boat had been low on gas and had had to return to port to
refuel. We had to wait an hour before she showed up again
to pick up the pilot.

We approached our first location in the early afternoon of
the fourteenth. The first borehole was to be spudded on the
northern slope of a submarine bank known as the Gorringe
Bank. This submarine mountain rose some 2,500 fathoms
(1 fathom = 1.8 meters) from the abyssal plain to a height of
about 100 fathoms below sea level (fig. 22). The area had
been previously surveyed by the French vessel, *Charcot*, of
CNEXO (the French National Center for the Exploitation of

the Oceans), and after carefully studying the seismic profiles, we decided that the optimum location should be one at a depth of 890 fathoms on the upper shoulder of the bank. The strategy, then, was to approach from the north and to locate our site when our "precision depth recorder" (PDR), using the echo sounding principle, registered exactly 890 fathoms.

As we drew near the site, however, Ryan and I became increasingly aware of an impending crisis, for the PDR had been out of order since we left Lisbon. Pete Garrow, Leg 13's electronic technician, had tried all night to fix it without even finding the source of the trouble. Meanwhile Ken Fors-

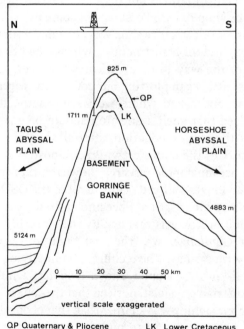

OP Quaternary & Pliocene   LK Lower Cretaceous

22. Sketch of Gorringe Bank and first borehole of the Leg 13 expedition. Courtesy of DSDP.

man, the marine technician, somewhat short on experience but eager to be helpful, had attempted a makeshift arrangement. He and Ryan had tried to hook up the old-fashioned fathometer to one of the two recorders on our seismic profilers. This fathometer could emit signals, but its registering equipment was completely out of date (we found out eventually that the registry was some 50 fathoms off in 900-fathom-deep water). If the fathometer signals could have been relayed to a modern recorder, however, then we would have had a substitute PDR.

Our ETA on site was 1800 hours. Around 1600 hours I went up to the electronics laboratory. But Ryan, who was to do the navigation, was not there. Reportedly he had gone down to the shop to help Forsman get some parts. The drafting table in the electronics shop was cluttered with resistors, condensors, and an assortment of wires; I had to push the junk out of the way in order to spread out the navigation charts. Just as I began plotting our course, the captain stuck his head in and asked where we were exactly. It seemed ironic indeed that such a question should be posed by the captain of the ship; navigation, after all, had been the traditional duty of the ship's crew. Precise positioning was so crucial in oceanographic research, however, that traditional methods no longer sufficed. Thus in all of the DSDP cruises satellite navigation (sat-nav) became an indispensable tool, and this equipment was manned by the scientific crew.

Glomar Challenger was the first "non-military" vessel equipped with sat-nav. The technique's underlying principle is the well-known Doppler effect. Doppler, a nineteenth-century Austrian physicist, noticed that the tone of a whistle of a train approaching was sharper than that of one departing. Stated in more scientific terms, the frequency (tone) of a received wave signal (whistle) depends upon the relative velocity between the emitter (train) and the receiver (ear). In sat-nav the emitter is one of the several artificial satellites

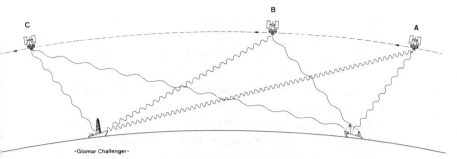

23. Diagram of satellite-navigation principles, illustrating changes in frequency of the signals received by vessels as a satellite passes from position A to C.

circling the globe, and the receiver is the moving ship. Comparing the known frequency of the signal emitted by the satellite with that actually received, one can compute the relative velocity between the satellite and ship, and the distances between them at several stages of the satellite's overhead flight. Since the path of the satellite is known precisely, the computed distances permit a determination of the ship's location by triangulation (fig. 23).

The sat-nav method requires electronic equipment for the reception of radio signals emitted by the satellites and a computer large enough to perform the complicated calculations. As the satellite and shipboard computers were relatively new inventions, precise positioning used to be the biggest headache for oceanographers, as Bill Menard described so vividly in his *Anatomy of an Expedition*. Being equipped with the sat-nav, we had no such headaches. That is to say, we should have had no such headaches if everything had functioned properly.

But Captain Clarke's inquiry prompted me to go down to the science office to find out why the latest sat-fix (fixed position by satellite navigation) had not been relayed to the ship's crew. Pete Garrow was down there. This was his third cruise aboard *Challenger*. An energetic young man, Garrow surprised me at Lisbon when he began to count "the days to go." We all did that near the end of a cruise, but not at the beginning before we even sailed out. Eventually I learned that Garrow was newly married and had come directly to Lisbon from his honeymoon trip to Denmark. It was thus understandable that he was less than eager just *then* to spend two months at sea. At that moment Garrow seemed to be at the end of his wits. Not only was the PDR kaput, but the sat-nav was malfunctioning as well. He had been working frantically to get things in working order since we left Lisbon, even forgoing sleep, and he had achieved no visible success. He was reprogramming the computer when I asked why the sat-fix had not been sent to the captain, and his answer was short and short-tempered: "No sat-fixes were relayed to the captain because we ain't got none."

So as the *Challenger* steamed relentlessly toward her destination, we were blind and deaf. The air gun, the acoustic source of our CSP, was still functioning, but the seismic record was terrible.

At 1630 the third mate, resorting to classical methods of "dead reckoning," told us that we were pretty near where we wanted to be; we had perhaps only a few miles to go. Ryan came up. Forsman's effort to hook the fathometer to the recorder had proved a complete failure. At this moment of near despair we were rescued by Garrow, who had finally worked out the computer programming; he came up with a sat-fix at 1647 hours that put us about four and a half kilometers due north of our projected location. It was now 1705, and we had actually overshot the mark. Turning sharply from a 226° course to 170°, we steamed to our rendezvous without PDR. We had to switch on the old fathometer.

Locating the site was only half of the problem. The other was positioning an 11,000-ton ship at a precise point. We could not simply bring the ship to a stop after we arrived on site. *Challenger*, like many other oceanographical vessels, towed several strings of gear: the magnetometer ("maggie") and two sets of air gun and hydrophones ("eels"), which extended about a kilometer behind the ship. These had to be pulled in before the ship slowed down and stopped. Yet the gear was needed for surveying; so it could be taken in only after a site was located. An easy way to solve the dilemma was to drop a sonic beacon when we passed over the site. But a beacon dumped out of a ship moving at 8 knots might not settle on the ocean bottom properly, and the risk of losing a beacon, which cost several thousand dollars, was very high. We were advised not to try this procedure except as a last resort. Throughout our cruise we developed a more primitive way of marking the site: we dropped floating markers overboard. Unfortunately these markers would drift with currents and were often two or more kilometers from the desired location when the vessel returned after we had secured her gear.

Nevertheless we tried the cheaper method at Gorringe Bank. The crew threw a red barrel overboard along with a red-flagged buoy when we identified our location from the record of the seismic profiles. Although the fathometer read only 835 fathoms, we calculated that we were some 890 fathoms above the bottom. We steamed southward with reduced speed as the technicians pulled in the "maggie" and "eels." Then we headed back to our destination; but the barrel had drifted about a kilometer away from the buoy. Using our old-fashioned fathometer as a guide, we found the buoy the more reliable of the two. When the fathometer was about to register 835 fathoms again, we notified the captain, who in turn relayed the message "all-stop" down to the engine room.

Once the *Challenger* was on site, we put into effect the dynamic positioning system to keep her stationary. A beacon, dropped overboard, settled at a designated spot on the

24. Roughnecks assembling the drill string. Courtesy of DSDP.

ocean bottom (fig. 12) and began to emit acoustic signals that were picked up by the ship's computer. When the ship drifted away from the signal, the computer would send a message to one or more of the ship's four side thrusters, which would bring the ship back into its original position. The system did allow a drift of about sixty meters, so the drill string, which could never be made strong enough to hold the ship anyway, could not be rigid. Several segments of the pipe, the so-called bumper-subs, were thus constructed like automobile shafts; they could absorb vertical motion yet at the same time transmit torques.

At 1800 the beacon was dropped; it reached the bottom ten minutes later. Meanwhile, the drilling crew began to assemble the drill string (fig. 24). They had to know the depth at this point, because they had to lower the drill string more slowly when it was getting close to the bottom in order to avoid a sudden impact that might break the string. Unfortunately, the CSP had recorded a depth of 890 fathoms, while the fathometer reading was only 835 fathoms. To be safe we gave the lesser of the two figures, namely 1,600 meters. As this depth was approached, the crew slowed down and very carefully let the string slip through the moon pool. But the 1,600-meter mark was soon passed, and no "touch down" was registered by the instrument panel. They passed 1,605, 1,623, 1,641, 1,659. . . . Each time a double segment of the pipe (18 meters long) was connected and dropped down, we would get dirty looks from the driller, the toolpusher, the drilling superintendent, and the operations manager. It was not the best way for two young and inexperienced co-chief scientists to win the confidence of their drilling crew. My apologies about the unreliability of our old-fashioned fathometer did not make any great impression.

So the embarrassment continued and the credibility gap widened. We passed 1,673, 1,691. . . . Still more dirty looks. Finally, I decided to return to the electronics lab to check the records again. I ascertained that the water depth registered by the seismic profiler could be as deep as 1,716 meters. As I was returning to the rig floor via the bridge I noticed a halt in the lowering of the drill string. Proceeding then to the driller's shack, I was relieved to learn that we had finally touched the bottom at 1,711 meters below sea level, almost exactly the figure given by our CSP record.

The main reason for our drilling on the Gorringe Bank was to compare the genesis of the Mediterranean to that of the Atlantic. Most earth scientists by this time believed that the Atlantic was opened when Europe and Africa drifted away from the Americas about 150 million years ago (fig. 25). The classic theory was first championed by a German meteorologist, Alfred Wegener, in the 1920s. It was reduced almost to oblivion for some forty years before it was ex-

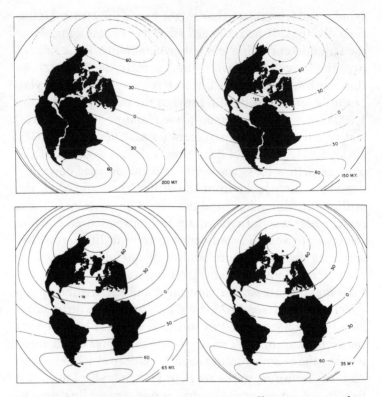

25. Stages of continental drift, starting 200 million years ago, when Europe, Africa, and the Americas were all part of the same continent, and ending with Africa and Europe continuing to move away from the Americas but approaching each other. Courtesy of the Geological Society of America.

humed by Vine and Matthews in the 1960s and reincarnated in a new form through geophysical and oceanographical discoveries made during the preceding decade. The deep-sea drilling cruise to the South Atlantic did much to convince obstinate skeptics like me who could not fancy continents traveling thousands of miles over the course of 100 million

years. By the time of the Leg 13 expedition the general principle of continental drift was established and we could apply
the principle to interpret the geology of the Mediterranean
and of its predecessor, the Tethys.

The sea floor spreading theory postulated that Europe, Africa, and the Americas were all part of a supercontinent, or
the one world, called Pangea, before it started to break up
some 200 million years ago. The Tethys, as the ocean opening up between Africa and Europe came to be called, was
born at the same time the central Atlantic opened up, when
Africa split away from Pangea some 150 million years ago
and Europe was still closely bound to the Americas. If this
hypothesis were correct, the Atlantic Ocean should be underlaid by a rock sequence resembling that of the Tethys:
the first sediments from the oldest part of the Atlantic and
the first oceanic sediments from the Tethys should be about
the same age. But the Tethys, once a mighty ocean, had been
largely eliminated by earth-quaking, mountain-building
movements. Its former grandeur had only been discerned by
studying the rocks in the Alps, which had been the sediments of the Tethys Ocean.

The Tethyan sediments were compressed and uplifted to
form the Alps when Europe eventually also split away from
the Americas, raced eastward, and caught Africa in collision.
Rocks forming the Tethyan ocean bottom are exposed on
the flanks of lofty Alpine peaks. These rocks have been called
ophiolites (snake rocks) because of their speckled green colors.
The ophiolites include basalt, which consists of cooled submarine lavas; gabbro, which has the same composition as
basalt but has coarser crystals; and serpentinite, which is
derived from rocks once deeply buried beneath the earth's
crust. The Tethyan ophiolites in the Alps underlie oceanic
sediments more than 130 million years old. Our drilling goal
on the Gorringe Bank was thus to drill through the ocean
sediments and see if we could find a 130-million-year-old
ophiolite beneath. It was a tall order, because previous ex-

perience told us that we might encounter very hard rock (chert) on our way down. If so, our drill bit might be completely worn out before we could reach basement.

We started coring operations in the small hours of August 15 and waited impatiently until the first core came up after dawn. It would seem to be a difficult trick to haul up a sample of the sea floor sediments from thousands of meters depth, but the principles involved are actually quite simple. The holes are bored by the rotary motion of the drill stem. At the depth where a core is wanted, the rotation is temporarily suspended and the drill stem disconnected at the derrick. A core barrel is then sent down to the end of the drill string just above the drill bit. After the barrel is properly seated, the drill stem is again connected and the rotary motion continues. The drill bit then bores deeper, while the pressure for circulating fluids is reduced so that the sediment is not washed away but enters a plastic liner inside the core barrel. After nine meters are drilled, the core barrel should be full. The rotation is again suspended, and a "sandline," like the line from a fishing rod, is sent down. At the end of the line is a hook, which should catch the top of the core barrel and pull it up to the rig floor. Once up the drill stem is disconnected again and the core barrel taken out (fig. 26). The next step is either to send another core barrel down for "continuous coring" or to drill ahead without a core barrel until the top of the next interval is reached where coring is desired.

Naturally many things can go wrong, and we did have problems during our Leg 13 cruise. Once, for example, an orifice in the core barrel that permits the water within the barrel to be displaced by sediments was accidentally closed. Then the hook of our sandline brought up only a barrel of water. Another time the sediments were too soft and dropped out of the catching device (the core catcher) as the core barrel

26. Two roughnecks removing a core barrel from the drill string.

was being raised. Yet another time the pump pressure was too high and all the sediment was washed away.

But if everything went as intended, the core barrel would be taken out of the drill stem, and the roughnecks would line up and pull out the inner plastic liner containing the core (fig. 27). The liner would then be laid on deck outside the core lab, where the marine technicians would cut the nine-meter core into six segments or "sections" and label them numerically. These sections would then be processed by Jenny Lort or by the technicians to determine the porosity, the water content, the gamma ray activity, and so on. Then the liners would be split into halves (fig. 28). One half would be retained as the "archives." The other would be the "working half." Shipboard sedimentologists would then take over. They would perform routine tests to determine the hardness of the sediments and describe the color, texture, structure, and mineral composition of the sediments. They would also put samples from the working half into small

27. Plastic liner from inside the core barrel; Maria Cita collecting ooze adhering to the liner. None of the costly samples from the deep sea was ever wasted.

vials for their own study later on, or for investigation by their colleagues in laboratories on land.

While the archive halves would be sent down to be photographed, the working halves would be sent to the paleo lab to be processed. Cita normally would wash the sediments through a sieve and pick out the foraminifera from the washed and dried residue, with a tiny brush. Stradner would take some of the sediments and make a "smear slide," then examine the slide under a high-powered microscope to make taxonomic identification of nannofossils. In some samples one might find foraminifera that had lived on the ocean bottom. These were the "benthonic" organisms that would be picked out and studied by Wolf Maync.

Our first core was a Lower Miocene ooze, some 20 million years old. The drilling continued. After we drilled through another 200 meters, we cored Cretaceous rocks that were about 110 million years old. The sedimentary sequence we penetrated was similar to that in the Betic Cordillera of southern Spain. This called for optimism. Yet the drilling rate at this point slowed perceptibly.

After a day and a half the rock formations we encountered got harder and harder, and the drilling proceeded at a snail's pace. Having not slept a wink during the night of the fifteenth, I was very impatient indeed. So I went down to the paleo lab to confer with Cita. How long could we justify our digging here when the sirens of the Mediterranean were beckoning? Cita felt the same way. All her life she had worked on Mediterranean rocks, and she could hardly wait to see a sample from the bottom of this "Roman lake." Maync and Stradner expressed similar sentiments. We decided to give Ryan one last chance. I went to the driller's shack and told him to finish drilling the double segment before taking a last core.

The grinding continued. Each screeching sound was sweet music. After an early supper I finally went to bed and Ryan

28. Bill Ryan, the author, and Travis Rayborn (drilling superintendent) examining a core from the eastern Mediterranean. Courtesy of DSDP.

29. Ophiolite found at the bottom of the core hole at Site 120 on Gorringe Bank. This discovery confirmed our idea that the Alps once lay under an ocean like the central Atlantic.

took over. I had not closed my eyes for more than an hour before Ryan waked me up again. He turned on the light and waved the piece of rock in his hand.

"We hit basement."

"What?"

"We hit basement," he repeated excitedly, showing me the piece of core.

It was an ophiolite called gabbro (fig. 29), exactly the same kind of gabbro I had seen ten years earlier on a sheer cliff under the Allalinhorn in the Alps. This turned out to be a most significant find, confirming our idea that the genesis of the Tethys or the Alpine geosyncline was related to the opening of the central Atlantic. We were elated.

So we had a lucky beginning. The location could not have been better chosen, and we had the good fortune to steer onto the spot while all but blindfolded. We had not run into any hard chert formation. We had not broken the drill pipe. We had not lost the drill bit. And we had achieved our objective. On top of everything, and most important, Ryan and I had started to work as a team and we were gaining the confidence of our crew.

With the drill string pulled up, dissembled, and secured, the ship left Site 120 at 0300 on August 17, steaming eastward for the Strait of Gibraltar.

5

BACK TO GIBRALTAR?

—o—

WE ENTERED the Strait of Gibraltar in the early morning of the eighteenth. I worked all day on the shipboard reports; then Ryan spent all night trying to fix the seismic profiler. Afterward, it did indeed give a much better record. The next morning, August 19, Pautot and I got together to plan for the next site. We had expected Ryan to sleep late and hoped to surprise him by navigating the *Challenger* onto our next site while he slept. Just as we approached the site shortly after midday, however, Ryan came to join us. He brought a telegram from Bob Gilkey telling us that they had found Dumitrica.

Our next site was scheduled to penetrate the sedimentary sequence in the sea floor of the Alboran Basin to reach basement. We wanted to learn what the M-reflector was, and we hoped to determine the age of this part of the Mediterranean basin. The information from the drill cores here also might tell us if the western Mediterranean was indeed being rifted apart as some scientists had postulated. We arrived at Site 121, south of Malaga (fig. 2), in the early afternoon of the nineteenth. Concern for Dumitrica's whereabouts was lost in the excitement of approaching the new site, finishing the reports for the previous one, and planning for the next. But once the drill string was down and started grinding, I went to the paleo lab to chat with Cita. She casually mentioned the presence of siliceous microfossils in the sediments and, knowing that Dumitrica was an expert on those, asked if I knew where he was. Remembering Gilkey's telegram, I ran to the bridge to ask the captain if he knew the whereabouts of our Romanian colleague.

"I don't know. We haven't gotten any messages since the telegram you got yesterday."

"Have you heard from Scripps at all?"

"No, but the radio operator did say that Scripps was trying to get a message across to us. He tried to tape it but the tape recorder wouldn't work. Anyway, he will listen again at seven o'clock tonight."

I was anxious that we get the message and tried to convey to the captain the importance of the call. "It looks like we will need Dumitrica's expertise."

"Aha, doctor," the captain said, "I knew there must be something behind your certain interest."

I returned to the derrick and talked with Ryan about the trouble communicating with Scripps. We were both irritated at the radio operator's apparent lack of attention.

Ryan and I called a meeting of the scientific staff in the lounge that evening. We were only halfway through when the captain came in:

"You were right, the message this morning must have been about Dumitrica. But we still could not get it."

"What is the problem?"

"I wrote a letter to the authorities a long time ago saying that we would have difficulties communicating with Scripps from the Mediterranean, but no action was taken. Naturally I knew it would be costly to get the necessary crystals. But it is just very difficult now. I am not trying to defend my radio man; he is new on this job. But we simply cannot do much. We get our messages from Scripps through monitoring broadcasts by the navy from London. We simply don't have people monitoring twenty-four hours a day."

"How did we miss the message this morning?"

"It wasn't our fault. The message came via the MERCAST broadcast. They announced that they had messages for so and so ships. Then they read off a list of ships, and the messages were broadcast one by one. When it came our turn to get a message, there was interference. The whole thing was garbled and unintelligible.

"Now," the captain remembered his purpose in coming down, "we just got a message from the British Admiralty on the distress channel saying that there were two scientists, an American and a Romanian, looking for *Glomar Challenger*. We tried to find a channel to communicate with them on. But it was impossible. All the channels were jammed. We shouldn't use the distress channel, but this was the only channel on which they could reach us. We heard the message loud and clear, and we presume that it came from Gibraltar. It should be ready for you shortly."

The captain then left, but soon the radio man came in and brought the message that we had been expecting all day; it read:

master clark wncu suggest challenger prepare to berth in gibraltar or to send blue fox for rendezvous

The third mate then herded us upstairs, where we found the captain fuming: "Whoever sent that message must either be extremely ignorant or extremely selfish. They are asking the mountain to move to meet Mohammad! With regard to *Blue Fox*, I would say that it is impossible to have my men driving the small launch all night against the 30-knot head wind to Gibraltar to pick up a guy who should have gotten on board at Lisbon in the first place!"

Just then, the radio shot out another call: "Radio Gibraltar, Radio Gibraltar, calling WNCU, calling WNCU. Please give location, course, speed. . . ." Again they had broadcast through the distress channel. The captain rushed to the radio phone. He tried furiously but in vain to find a channel through which to reach our callers. Finally he sent a message telling them to wait for the requested information to be sent by commercial telegram.

I had been staying up late since we left Lisbon and thought I might catch up on some sleep the night of the nineteenth while Ryan stood watch. Ryan seemed to have sensed something, however, and kept me up until shortly after midnight, when we hit what seemed like hard rock. The core barrel jammed. We pulled it up soon enough and found that we had cored sandstone—not just the friable sandstone that breaks between your fingers, but a hard, cemented sandstone.

Roy Anderson, our operations manager, had the authority to stop the drilling if there was even the slightest chance we would encounter oil or gas. His orders were to halt drilling immediately if we got into a lithified formation at this site. We too were acutely aware of the danger of oil spills. If we hit a tight gas pocket, the consequences could be catastrophic. Faced with the threat of having to abandon our hole, we began to study our seismic record again, looking for an alternate site to reach the basement. There were indeed other possibilities, but that would mean the loss of a day or two, and time was gold to us, literally.

Before deciding to abandon the site, Ryan cut a thin section of the sandstone and I studied the slice under the microscope. It seemed that the rock had come from a concretionary layer in a soft formation, not from a hard formation that could seal an oil or gas reservoir. Anderson gave us a stay of execution, and we waited with suspense to see what the next barrel would bring up. There was an audible sigh of relief when we saw the soft green ooze in the next core. Anderson gave us the go-ahead. But Ryan and I decided to stand alert in case any new crisis should arise. Thus went another sleepless night. The drilling proceeded slowly.

In the early afternoon of the twentieth, a radio message finally came through from Gilkey. It confirmed that he and Dumitrica were in Gibraltar. They would come to us aboard a patrol boat of the Royal Air Force. Meanwhile a telegram also reached us, urging us to remain on location until they arrived. The captain talked on the phone with the Gibraltar Radio and was notified of their scheduled departure at 1500 hours.

The sun was shining but the wind was blowing hard and the sea was very choppy. We all went to our cabins and started writing home, for Gilkey could take mail ashore after he delivered Dumitrica. The chief mate estimated that with their fast patrol boat and the wind behind them they would be here in two or three hours. So the suspense continued. I decided to lie down and catch some sleep. I tried for a quarter of an hour. Useless. I got up, took out my binoculars, and went up to the bridge. Rumors were flying. Every little blip on the radar screen was construed by rumor-mongers to be the RAF patrol boat. At one stage of the game it was only six miles from us. There were almost as many false alarms as there were boats passing by us heading east through the strait, and there seemed to have been hundreds that afternoon.

By 1900 hours when it began to get dark, we began to lose hope that they would come. I retreated to the lounge to listen to a tape of Beethoven's Piano Concerto in E-flat, when our secretary and technicians came down from the bridge to summon me. From the way they were joking I had a foreboding of bad news.

Captain Clarke, Anderson, and Ryan were already on the bridge when I got there. They explained that we were ordered to return to Gibraltar to pick up Dumitrica. I exploded. Here we had worked days and nights trying to achieve our scientific objectives with maximum efficiency. Great tension and care had gone into choosing the sites; we were under

constant pressure to make the right judgment for drilling
and coring operations; we were always having to convince
our colleagues of the wisdom of our decisions and to pacify
them in their impatience while pursuing our ultimate objectives. All the days that we were trying to save hours, to
save minutes, so that our ship-time could be put to the most
efficient use of gathering scientific information. Now we
were ordered back to Gibraltar to pick up a guy whose as-
sistance we might not need. Even the loss of twelve hours
of ship-time would be equivalent to more than $10,000—
my whole year's budget for geologic field work back home!
"No," I said emphatically, "we are not going back."

We started to study the chain of command. Gilkey did not
have the authority to order us to turn around, but he claimed
to have the concurrence of Ken Brunot, the project's manager
and thus Anderson's boss. Although we did not yet have a
message from Brunot, we might still have to follow orders.

In this emotion-charged atmosphere Captain Clarke suggested a compromise. We would not return to Gibraltar;
instead, we would schedule a rendezvous with Dumitrica at
Malta. Ryan and I agreed, and we sent out a firm but dip-
lomatic message:

> dumitrica's presence desirable but not urgently required
> stop suggest alternate rendezvous at malta stop

Meanwhile, Anderson agreed that if we received no message
from Gilkey within twelve hours and if we should finish
drilling by then we would proceed eastward to our next location.

It was about 1830 when I got into my cabin, had a shower,
swallowed a shot of Johnny Walker, and climbed into my
bunk. Barely had I closed my eyes, when the light was again
turned on and Ryan rushed in with a sample in his hand:
"We reached basement."

It came as a surprise. We had progressed rapidly until the evening before; then the progress had slowed. During the next twenty-four hours we had had several false alarms. A number of times the driller told us that he had hit something hard. Then when we pulled up, we would find only a few chips of cemented sands, or a jammed core barrel. So I became very discouraged. We studied our seismic record and revised again and again our estimates, sometimes too optimistically, then too pessimistically. When I went to bed, I fully expected an eight-hour sleep. Yet the next barrel (the twenty-third core at this location) brought up the basement.

I took one look at the rock chip Ryan was wielding and decided it was basalt. Ryan said that they had already sent down the center bit to drill deeper but that we could recall the bit and core more to confirm our findings. I agreed to the latter course. So Ryan rushed back to the derrick, and I dressed and went to the core lab.

In fact, as we found out later, the rock was not a basalt but a metamorphic rock. In any case, we did hit basement. We waited another three hours for the last core to come up and had our final confirmation. Ryan, elated beyond words, invited a few people back to our cabin for a victory celebration. Then he hit the sack. But I was still too excited to sleep, so I went up to the bridge to talk with the radio officer about our communication problem.

The curtain to the second act of the Dumitrica drama fell on the evening of the twenty-first when we departed from our Alboran Basin site and were on the way to the Valencia Trough off Barcelona. The captain gave us a message from Brunot ordering us back to Gibraltar. Fortunately, however, the inefficient MERCAST system saved us from this command; the telegram came too late to affect our decision.

6

MURPHY'S LAW

——o——

ALTHOUGH the M-layer had been identified at Site 121, we were not certain at which depth we had penetrated the formation. We did recover samples from a hard-rock formation at 700 meters depth where the M-reflector should be. This rock was very fine grained, but our sedimentologists were not sure exactly what it was. Later on Nesteroff was to determine by X-ray diffraction that the rock was dolomite (a carbonate of calcium and magnesium), a chemical sediment formed when the Mediterranean dried up. But we did not know this then and cherished hopes of determining the nature of the M-reflector at this third site. Site 122 was situated in a submarine channel off the Barcelona coast, called the Valencia Trough (fig. 2). In this area of submarine erosion, the M-reflector was covered by a very thin sedimentary layer and was thus easily within reach. We approached the site from the southwest, at 062°, in the early hours of the twenty-third. It was almost impossible to get any sleep. I stayed up, first writing some reports, then joining the crew to watch a movie, a satire on James Bond. Around 0230 I went to the electronics lab to join Ryan as we steamed onto location. We were to drill at the bottom of the submarine valley. Since Pete Garrow had finally fixed the PDR, we foresaw no great difficulties in finding the place.

When I went up to the lab, everything seemed to be in excellent order. We approached the site slowly, crossing highs

and lows, and located the channel. We threw a buoy overboard as we crossed over our preselected site at 0346, hoping
the radio signals emitted might be strong enough to be picked
up by our radar screen. They were not, but it did not matter.
With the PDR we could find our valley easily on our way
back. So the technicians pulled in the gear—in a record time
of twelve minutes, as they proudly reported. The sat-fix told
us that we were on a parallel course but a little to the northwest of the surveying traverse by the French vessel *Charcot*.
Since our drill site was located on that profile, we asked the
captain to make a 180° turn to the right. We were now heading back to our location, course 239°. We watched the PDR
intensely. The water depth increased steadily as we returned
to the submarine valley. Eight minutes to go, seven, six,
five. . . . We notified the captain to slow the engine and prepare to stop. He did as requested. But the current began
playing tricks on us. As we reached "all-stop" over the station, we noticed that we were being driven southwards and
had overshot the axis of the channel. The PDR showed a
climbing bottom profile, up the southwest wall of the channel. We asked the captain to go astern. He relayed the message to the engine room. Still the bottom topography was
climbing, we were not backing up with sufficient speed to
counter the northeasterly current. At this situation we asked
the captain to turn completely around to head 030° again.
Our problem was complicated, however, when the turning
mechanism momentarily failed. Finally when that was fixed,
the gyrocompass registered a 300° course, because the captain thought this was the direction we had requested. We
thought that the steering gear was locked in that position
and could not be oriented to the desired direction. The vessel
continued to sail away from our target and the bottom topography continued to climb. Eventually, the misunderstanding was clarified and the captain turned the vessel to
030° at 1 knot speed. Still we were drifting away. We turned

again, to 060°, thus making a 180° turn from the original course. Yet the needle of our precision depth recorder continued to climb and we were driven even farther away. At this stage Ryan and I began to panic; we felt almost impotent to master the situation. The bottom topography seemed to rise in every direction, no matter whither the ship was headed. But soon we recovered our wits and discovered the culprit. We were steaming at 1 knot, and the current must have been moving at least 1.5 knots so that the ship was always drifting backward. We decided to maintain the 060° course but increased the speed to 2 knots. With a sigh of relief, we watched the PDR needle finally print a downhill curve. At 0540, more than two hours after first crossing over the site, we finally maneuvered the *Challenger* to a location directly above the axis of the channel.

After the beacon was dropped, Ryan went to bed. I decided to stay up until 0800, having made an arrangement with an acquaintance in Bern to contact him by ham radio every Saturday evening at 2000 and every Sunday morning at 0800. We tried for fifteen minutes but could not make any connection. So I too went to lie down. But I got up after three hours, being too keyed up to sleep.

Drilling started in the afternoon and proceeded smoothly enough. We penetrated the soft upper sediments at an unusually rapid rate, with cores hauled on deck every hour and a half. Things went on beautifully for awhile, so I told Ryan that I would have some supper and then catch some sleep. Shortly before midnight Ryan came in and turned on the light.

"Ken, I have news for you—some good news and some bad."

My mind being still preoccupied with the Dumitrica affair, I assumed that the bad news was an order for us to turn back to Gibraltar.

"Okay, shoot. What went wrong now?"

"The core barrel got stuck, the drill bit plugged, and the circulation failed. We had to get out of the hole."

Mechanical difficulties like this plagued us throughout the cruise. Sometimes we blamed our leg's being the unlucky thirteenth, but in fact the difficulties were unavoidable; our technical capacities would not allow us to drill through thick sandy or hard formations. In contrast to drilling operations on land, where drill cuttings (chips of hard rock cut off by the drill bit) are brought up to the rig floor by circulating muds, the cuttings of the sediments encountered in deep-sea drilling are brought up only to the sea floor, where they form a circular pile around the drill hole. As long as drilling continues and seawater is circulating to keep the hole free of debris, there is no problem. But as soon as the drill stem is disconnected and the circulation stopped, sands or cuttings from the debris pile fall back down into the drill hole. The drill stem is then buried in the sands (or cuttings), and the friction prevents its further rotation, thus further penetration by drilling is impossible. In some instances the sands have been known to enter the core barrel and the inside of the drill stem. Then the crew cannot even fish out the barrel. This is what happened on that night in the Valencia Trough. We did not have enough experience when we selected our site to realize that we should have avoided the sands in a submarine channel; it would have been preferable to drill a muddy section, even if it had been several times thicker, for muds tend to wash away and would not have formed a debris pile around the drill hole.

The jammed core barrel would mean a loss of twelve hours of drilling time—the time we would have had to spend picking up Dumitrica.

"Is there anything else they could do?" I asked.

"No, they are pulling up the drill string right now."

"What was so good then?"

"We found gypsum at the bottom of the Pliocene."

Previous seismic surveys had suggested that a layer of salt underlay the Mediterranean sea floor, but there was never direct proof. The age of this salt was also a controversial subject. A French professor of geology who was a member of the French Academy and the teacher of the two French scientists on board maintained that the salt was Triassic, or some 200 million years old. A few young rebels thought that the salt might be as young as late Miocene, or some 5 or 6 million years old.

We did not find salt at Site 122, only gypsum. But gypsum is an evaporite mineral, calcium sulfate ($CaSO_4 \cdot 2H_2O$). It must have come down when the seawater was turned into a brine by evaporation. Halite ($NaCl$), or rock salt, would have been precipitated if the seawater had been further concentrated. The discovery of gypsum, which was being dated as early Pliocene or late Miocene, was therefore the first concrete evidence we found for a Mediterranean salinity crisis.

I got dressed and went to the core lab. They were pulling up the drill string outside. Ryan and I looked at the five grams of washed residue that was all we had retrieved from the last core. Indeed there were gypsum crystals glistening among black grains of volcanic rocks. Ryan had thought that they were chips and fragments dug up from a bedded evaporite. I was disconcerted to find that the gypsum looked like detrital grains, however. The crystals were all of similar sizes and about as large as the grains of the volcanics, which could be identified as an andesite. Since the latter were obviously detrital, it would be difficult to convince anyone, including me, that the gypsum grains did not originate the same way. Ryan knew of gypsum deposits in Spain nearby. It seemed quite likely indeed that these grains were derived from the Costa Blanca. Slowly our excitement subsided and disappointment crept in; perhaps we had found nothing unusual, just some gravelly sand of gypsum and volcanics, some debris that had been washed into this deep channel. Feeling much discouraged, we went back to our cabin, leaving word that we should be notified when the core barrel reached the derrick floor.

In a couple of hours I was awakened by Roy Anderson. I went to the derrick floor and found the drilling crew still trying to pull the jammed core barrel out of the drill collar; mainly they had to wash the sands away. After watching for half an hour in complete frustration, I entered the core lab to find something more useful to do. There I met Anderson. He was pessimistic. Since we might get stuck again if we reentered the hole, he advised that we look for a new location nearby.

I went back to the cabin to consult Ryan. Only half awake, he was much annoyed. This time, I suggested that we quit the hole. In the matter of drilling we had to defer to the judgment of the operations manager. It was a bitter pill to swallow, but glumly we went together to the core lab and studied the charts for an alternative location. I kept my temper, and Ryan tried to keep his. Actually things did not look too bad. An alternate site about fourteen miles to the north seemed promising. There we should be able to avoid the sands that had caused the pipe to bind here.

A new site was thus selected and its coordinates determined. I went to the bridge and talked to the third mate on watch. Since we had to move on to the next location, the captain was awakened. Soon the drill string was dissembled and secured, and at 0622, August 24, we departed for Site 123.

This sudden change in plans brought forth new confusion. The technician on watch was ordered to make up a new buoy. But he also had to stand by the sat-nav to get the latest fix. As we steamed toward the next location, the captain asked for sat-nav results, and got the message that no fix had been obtained. Ryan's temper flared. He rushed down to the science office where the sat-nav set was and chewed out Jones, the technician.

Jones was actually one of our most courteous and helpful
assistants. Except for a Beatle haircut, one would have thought
him a square. He and I had served together on the Leg 3
cruise across the South Atlantic, and I had greatly appreci-
ated his thoughtfulness. On that morning of confusion, however, he had found himself alone on watch. The lab officer,
who as foreman of the technicians would ordinarily have
been up and around on such occasions, had been bedded
down with the flu for two days. So Jones found himself torn
between two equally urgent tasks. At first he had thought
the ship was still maneuvering at a variable speed, when a
satisfactory sat-fix result cannot be obtained. So he decided
to attend to the job of painting the buoy, despite our repeated
instructions never to cut out a sat-fix when we were approaching a drilling location. Meaning well, he had simply
made a mistake. When Ryan went down, it was late and we
had already lost the chance of getting a fix. Nevertheless,
we had figured out that we would be crossing over the location in a few minutes, at which time we should drop the
buoy. But Jones, just given hell for missing the last sat-fix,
decided to stand by for the next, which incidentally would
not come in for another hour. So when the captain yelled
"let her go," nobody was around to execute the order. Furiously Ryan rushed down to the fantail of the ship himself
and pushed the buoy abroad. When he came back, I was
shocked to see "blood" on his hands. It was only fresh paint,
however; Jones had just finished painting the marker.

We slowed down the engine and turned back to the location marked by the buoy. This new site, like the previous
site, was situated over a submarine channel, but this channel
was not as deep as the Valencia Trough. Furthermore, we
were to spud the hole on a shoulder at the southwest wall
and thus were hoping to avoid the channel sands. The dawn
finally came, and we started the new day at this new location.

7

A DEPOSIT OF UNUSUAL
GRAVELS

—◯—

ON THE MORNING of August 24 both Bill Ryan and I felt
dead tired. The crew was lowering the drill string and there
was nothing for us to do, so we should really have gone to
bed, but we were too nervous and depressed. More than ten
days had gone by and the ship-time alone had cost a quarter
of a million dollars, but we had not even solved the simple
puzzle of the identity of the M-reflector. Silently we stayed
in the core lab, Ryan mechanically washing and sorting out
pea gravels from the bucket of sands brought in the night
before and me sitting on a high stool watching. As the morning wore on and the collection grew, I became more and
more amazed by what I saw. There was no trace of anything
that could be positively identified as coming from the nearby
continent. The gravels could not have been brought out here
by turbidity current from the Spanish Coast. What was the
meaning of this unusual gravel?

Toward the evening of the twenty-fourth the drill bit that
had penetrated the last hole was brought in. Caught in the
teeth of the bit, which had been stuck near the top of the
hard layer, were fine aggregates of bedded gypsum. So we
had proof that the gypsum grains in the volcanic gravel had
come from the erosion of an older gypsum layer, the inorganic residue left behind by evaporated brines. Ancient eva-

porites are commonly lithified (converted into rock). Thus we had the first solution to our mysteries: the M-reflector, or the hard layer underlying the Mediterranean, was a late Miocene evaporite.

The new site was positioned over a buried submarine mountain (fig. 2). Since the seismic records showed that the M-layer was absent here, we had planned to drill through a thin sedimentary cover to reach basement. Our first core, a mud sample, was an ice-age deposit and showed nothing particularly alarming. When the driller cut the second core, however, we noticed that the speed of penetration was notably faster. And when the core was hauled on deck, we found again sandy gravels. It was not entirely surprising to find gravels, as we were on the edge of a submarine channel. We were sure that the gravels did not constitute the M-layer, which should be absent. Nevertheless, Roy Anderson, the operations manager, was rather nervous after the unfortunate experience we had had the night before. He had visions of the drill pipe being stuck in a gravel again and of losing the whole drill string.

I opted for the risk, for neither our own nor the *Charcot* seismic records registered the presence of a reflecting horizon that could be the top of a thick gravel deposit. Ryan wanted to go along with Anderson, however, being afraid to lose the drill string. Into this stalemate came the captain with a suggestion. Theoretically it was not his duty to worry about the technical details of drilling operations, but he expressed the opinion that calculated risks could be justified. If the information we hoped to obtain was a sufficiently important objective, we should go ahead. Furthermore, even if we should get stuck again, we could blast our way out of the hole by leaving the bottom-hole assembly (i.e., the lower part of the drill string) in the hole and retrieving the rest of the drill pipe (more than two kilometers long). The loss would then be around $20,000; but a day of ship-time cost more. And after all, he said, we came out here for scientific exploration, not to establish a record of minimum equipment loss. The captain's advice encouraged us. To our questions about equipment losses on other DSDP cruises he replied, "Oh, everybody lost bottom-hole assemblies; Leg 6 lost nine."

Ryan and I talked the matter over once more. He was now convinced that we were not likely to encounter a situation like the one we had yesterday, and it would simply be a waste of ship-time if we continued to fool around looking for still another location. Either we should take a chance here or we might as well just quit and do something else at another spot altogether.

Anderson finally relented. He would wait at least until the next core came up. If it did not appear that we had a massive gravel deposit, he would let us go on. Just then the ship's public address system notified us of the arrival of the third core. Anderson rushed to the drill rig. Ryan and I followed and tried to make up our minds whether to give in and quit if we had cored gravels again. It had cost us a day's ship-time on location so far and we had not yet found anything. Quitting now would mean a net loss of $25,000. If we continued to drill, however, we might soon hit another formation. There was less than a fifty percent chance of losing $20,000 worth of equipment, so we decided to take the risk. But the difficult task of persuading Anderson remained. Fortunately we did not have to go through this, for the third core contained only mud. Anderson now felt that the risk of further drilling was justified. So we could proceed and save the hole.

Roughnecks brought the mud into the core lab from the derrick floor for testing. Never was there mud that gave Ryan and me so much joy. The sedimentologists, not appreciating the gravity of the situation, joked about the chance of hitting an oil gusher. We two did not relish the humor just then;

Anderson might take such jokes seriously. Meanwhile, Cita's voice came through the public address system to tell us that the mud was Pliocene. So now we knew we were no longer drilling in the ice-age deposits, which commonly contain gravels. After relaying the message to Anderson to pacify him still further, I decided to catch some sleep.

After a couple of hours I was awakened by radio noises. All day long there had been conversations between the captain and the Global Marine agents in Spain. Dumitrica was supposed to arrive at 2000, then at 2200, and finally at 0030 hours. From Barcelona, from Las Palmas? Nobody seemed to have a very clear idea. This time I did not want to stay up. I had begun to wonder if Dumitrica actually existed and would not believe that he was really coming until I saw him with my own two eyes. At about 2215 there was an urgent radio conversation filtering out of Karl Wells' cabin, which was adjacent to ours. Thinking that Dumitrica must have come, I got dressed and went up to the bridge. Several other people were already waiting there. The captain was talking again on the radio phone: "WNCB, WNCB . . . Whisky Night Charlie Bravo. Whisky Night Charlie Bravo, calling. . . ."

The radio message from the other party was badly garbled. We saw some blips on the radar screen, bearing 090°, but the captain rushed to the chart room and yelled that the launch was somewhere to the west of us, bearing 275°. He told us that it was hopeless to use a radar scope to search for a small wooden vessel, some sixty-five meters long. They would find us much more easily than we could find them. Then he ordered a seaman to send out light signals for the approaching boat.

Meanwhile, looking toward the derrick, I noticed that they were bringing up another core. I hurried to the core lab. They had brought in some four and a half meters of stiff clay. Our gamble had paid off. We were definitely in safe ground now. Ryan continued to stand watch on the derrick floor, while I wandered back to the bridge. There was still no sign of the launch, but she should have been here an hour ago if she had kept her heading and speed. I was a little worried. The captain laughed. I was becoming well known as the pessimist on board.

After midnight another core came up. When I returned to the bridge they told me that the launch was now in sight. I went out, and indeed there was a light like a star on the horizon. So that was the boat that would bring Dumitrica. On the radar screen one could now follow her progress. Four miles, three miles, two miles. . . . I went to fetch my field glasses. Soon I was able to identify Bob Gilkey with a dark-haired young man at his side. They were riding in a Spanish fishing boat, which was soon alongside the *Challenger*. First they passed over their briefcases and baggage. Then the captain threw a life jacket down to Dumitrica. Dumitrica did not want to wear it, however, and was not persuaded by Gilkey's insistence. We helped them on board; and finally we had Dumitrica.

The newcomers were led to the mess hall, where Gilkey recounted his saga. He had chased Dumitrica through Paris, London, and Tangiers before finding him as a *persona non grata* in the care of the wife of the United States naval attaché at Gibraltar. He had found him twelve hours before we sailed past Gibraltar, but they could not communicate with us. At the zenith of his frustration, Gilkey had sent us a message asking us to return. Eventually, however, they secured the Spanish visa, flew to Tangiers, then to Barcelona; then they drove to a local fishing village and bounced for thirteen hours on a fishing boat to catch the *Challenger*.

The whole episode started with the harmless negligence of a girl in a telegraph office in Paris. Dumitrica had given his Paris address when he wired Scripps for help. But the girl for some reason did not include that address when she sent out the telegram. After waiting in vain for a week in Paris, Dumitrica returned to Romania, telephoned Peterson at Scripps, who in turn contacted Gilkey in Lisbon. Dumitrica was then supposed to meet Gilkey at Le Bourget Airport in Paris; but Gilkey's plane from Lisbon was late, and the traffic from Orly to Le Bourget caused further delays. Dumitrica was on his way to London when Gilkey tried to page him in the Paris airport. By the time Gilkey got to London, Dumitrica was gone again. Through the good office of the British Government (Scotland Yard), Gilkey found a record indicating that Dumitrica had flown to Tangiers and Gibraltar. Gilkey could not get a reservation on the flight, but he managed to make the trip on standby. After he reached Gibraltar, he received a message to call Mrs. Sapp, and he found Dumitrica in her care.

As we talked, another core was hauled in. This one consisted of green volcanic debris. It seemed that we had hit a volcanic ash layer; these are commonly intercalated in the young sediments of the Mediterranean. Ryan and I were both excited by this discovery. Soon I persuaded Ryan to go to bed while I continued "well sitting." He was happy to go, mentioning his nervousness during the last several hours. The way the drill pipe went down made him constantly scared of having the string stuck again.

After he went to bed, I went to the rig floor to ask them to take in another core after drilling thirty meters further. It was not long before the core barrel was on deck, but it was a bad omen when we found the core catcher empty. My heart sank when a roughneck looked through the core barrel, confirming that it was empty too. So it was last night's story again. As I walked to the driller's shack, my worst fear was confirmed. We were stuck. We tried to work the drill string up and down. The normal weight of the string was 25,000 lbs., and we exerted up to 40,000 lbs. tension; but the string stretched and would not be budged. Mud was poured in to increase the efficiency of the pumping, but it did not help.

I was tired and discouraged. Around 0530 I decided there was nothing I could do by staying up. They would have to blast the bottom-hole assembly away. Not wishing to be tortured any longer, I went to bed.

8

EXECUTION STAYED

—o—

AT ABOUT six o'clock in the morning of August 25, half an hour after I had gone to bed, Roy Anderson came in to tell us that the pipe was still stuck and that pumping with mud did not help. They would try some more, but then they would resort to blasting. This came as no surprise to me. Although equipment loss was a part of the game, still I hated to think that we would lose $20,000 and had very little to show for the hole.

Around nine o'clock, still half awake, I realized that the discovery of the volcanic ash bed was a significant finding and decided that we should rig up the air gun and pass over the station again to get a good seismic record before we departed. We might be able to trace the ash layer, which was the culprit of our trouble, across the Balearic Basin. I tried to awaken Ryan, but he was too sleepy for any decision making, so I got up alone and gave instructions to the bridge and to the technicians to get the seismic equipment ready. Then I went to the rig floor where I saw that the crew was lowering the charges. With a sinking heart, I returned to the cabin, only somewhat consoled by the thought that nothing would prevent me from getting an eight-hour sleep. But at eleven, Nesteroff came in and told us that they had managed to pull the drill pipe free. So we had to decide again if we would abandon the hole or drill ahead.

Apparently Karl Wells, the ship's electrical engineer, had

been just about to detonate the charge when the drill string was pulled free. The crew started to dissemble the string before Wells went to Nesteroff and told him to check with us. Ryan and I went immediately to the driller's shack to discuss the situation with Travis Rayborn, the drilling superintendent, and then with Anderson. After investing one whole day on the hole and given a second chance, we decided to push our luck and drill further—or as Anderson said, "go for broke."

We drilled another hundred meters and pulled in a core. Again we found green volcanic ash. We now realized that the ash did not constitute a thin layer intercalated in a sedimentary sequence. The volcanic ash was a thick deposit on the flank of an extinct submarine volcano. We had in fact drilled through the sedimentary section before we first encountered the ash. Now we had penetrated some 150 meters of volcanic debris. It would be nice to find solid lava-flow rocks further down. But we might push our luck too far and get stuck again. With regret, we thus decided to pull out and cement the hole.

But once more our incredible luck had held. Instead of failure, we met huge success, hitting basement just there and collecting enough samples to allow our colleagues on land eventually to determine its age. They concluded that the ash and the underlying volcanic rock were about 20 million years old. There had been a chain of volcanoes here then, which had spat out ash that had accumulated to layers hundreds of meters thick. Later the Valencia Trough was submerged under a deep sea, and skeletons of floating animals and plants were sedimented to form oozes. The Mediterranean eventually dried up, and layers of evaporites were deposited over the whole basin. Gravels and sands were transported by rivers draining the newly emerged volcanic mountains, forming the gravelly deposits we had found at the previous site. Finally, at the beginning of Pliocene, some

5 million years ago, the Strait of Gibraltar opened and the sea came back; and deep-sea oozes were again deposited in the flooded Mediterranean basins.

I asked Rayborn later how we had managed to free the drill string. It had been almost a miracle, he said; nothing like it ever occurred before in all the years of his long drilling experience. The dynamite charges had been sent down. Wells was at the other end, ready to push the button. The last step was to raise the drill string to give it some tension. So the string was drawn tight, and the tension was increased from 25,000 to 40,000 lbs. But just when he was about to give the signal for blasting, he noticed that the needle of the tension gauge was creeping downward. It continued to fall until the dial finally showed the normal weight. Suddenly he realized that the pipe had worked itself free. He frantically signaled the blasting crew to stop operations, barely managing to make himself understood before the button was to be pushed. When I asked him how the pipe did finally get free, he replied with his typically wry humor, and in language worthy of a roughneck: "When we sent the dynamite down, it scared the —— out of her. So she decided to get the hell out of that damned rat hole."

9

ALL'S WELL THAT
ENDS WELL

—o—

WE MOVED on to our next drill site south of Mallorca in the Balearic Basin (fig. 2) on August 26, our objective being to determine the age of this basin by taking samples of the basement rock. Two theories on the origin of the western Mediterranean gave us our starting point. I have mentioned the theory of Emil Argand, who proposed in 1922 that Corsica and Sardinia had orginally been attached to the coast of southern France and Spain (fig. 30); these islands supposedly drifted to their present position many millions of years ago, leaving a deep pit that was partially filled by basaltic lava from the interior of the earth. This pit constituted the Balearic Basin. An alternative theory postulated that a part of the continent between the Balearic Islands and Corso-Sardinia sank to oceanic depth for some unknown reason, thus giving rise to the Balearic Sea. Evidence on land favored Argand's hypothesis, but it was not unequivocal. We thought we might resolve this controversy if we could recover a piece of the hard rock basement underlying the soft sediments of the Balearic sea floor.

Our strategy had been to locate our site directly over a spot where the sediments above the basement were relatively thin so that we might drill through the sedimentary veneer more easily to reach bedrock. We had chosen one

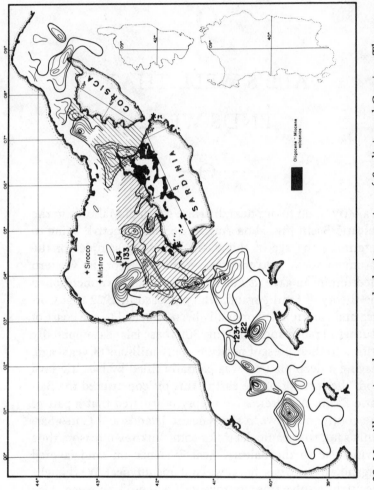

30. Map illustrating the counterclockwise rotation of Sardinia and Corsica. The contours show the intensity of magnetic anomalies of submarine volcanoes buried under soft sediments.

such location south of the Balearic Islands, which had been surveyed by the French vessel *Charcot* in the spring before our cruise.

Sailing to our destination from the northwest, we noticed all morning long that our course tended to drift off to the west drawn by winds and currents. We planned, therefore, to head for a point due east of our selected site, then turn sharply to the west and pass over the site, dropping the buoy on the run and pulling in the gear before making our way back to the station. We could then approach the target from the west, countering the current. If we should overshoot, as we had at Site 122, it would be a simple task just to back the vessel up with the drift. The key to our success would be to obtain precise locationing during our approach. Satellite navigation was a must, for classical methods of dead reckoning were simply not accurate enough. If we should miss our preselected site by half a mile, we might fail altogether to achieve our objective.

We steered a 114° course toward a point east of our goal, expecting a sat-fix to come in at 1306. Since the ship had to maintain a constant course and speed when satellite navigation data were being gathered, the captain decided shortly before 1300 hours to order a reduction in speed from 9 to 6 knots so as not to overshoot the turning point. Alas, the satellite appeared somewhat earlier than scheduled, and we were already tied up with a sat-fix before the captain issued his orders. Now we had to wait until the computations were finished in the science office. We waited for twenty minutes, however, and were still not getting the fix. So Ryan went down to see what was going on. He was furious to find the marine technician on duty reading an old newspaper between incoming signals, when concentrated attention was absolutely required. To make the matter worse, the young man had missed the significance of several signals and had made stupid mistakes in his computations. The captain had to wait for the figures before executing the turns, but the computation for this particularly important sat-fix took three times longer than usual. We did not obtain the results for the 1300 fix until 1340. We watched helplessly as we overshot our turning point. The game plan was spoiled and the arrival time delayed. This delay caused further uncertainties in our navigation.

A somewhat comic interlude occurred just then. Forese Wezel was sunning himself on deck when he saw a red flag on a buoy in the water, the signal that we were approaching a drill site. He ran into the electronics lab just as Ryan went down to chew out the technician on sat-nav watch to ask if we were already on station.

"No, we haven't made the turn yet," I told him.

"But I just saw the buoy floating past me."

He had barely finished his sentence when Dell Cover, an eager and well-intentioned technician rushed in, bringing the message that they had lost the buoy. Ted Gustafson, our lab officer, got upset:

"Did you guys secure it?"

"No, we forgot, and the wind blew it away."

"For crying out loud, go make another one!"

Sheepishly, the technician went out as ordered. Trying to pacify Gustafson, I assured him that we still had an hour to make another buoy.

While tempers flared, *Glomar Challenger* groped her way onto location at 1600 with the aid of dead reckoning. We could not be certain of our precise whereabouts. But I was not at all eager to drift along and wait for the 1705 sat-fix to confirm our location. We had already lost too much time in the confusion. So we decided to drop the beacon and stay where we were. As it turned out, we were two full kilometers off, and the sedimentary sequence at Site 124 was more than twice as thick as that at our preselected location. We spent three days drilling through 400 meters of sediments without getting anywhere close to the basement. Thus we were unable to achieve our original objective. All's well that ends well, however. Stumbling onto the wrong location, we obtained a beautiful suite of cores of the Mediterranean evaporites, which told us a great deal about the fascinating time when the Mediterranean had dried up. If we had maneuvered the vessel onto precisely the point we had wanted to go, we might have missed most of the evaporites section. We probably could have reached basement, as we later did in Hole 134 on the other side of the Balearic, but the basement sample would probably not have solved our problem as we had hoped.

Things were settling down to routines by the evening of the twenty-sixth. Outside on the rig floor, the roughnecks were putting the drill string together. In the science lounge, we watched a movie depicting the last days of the British Empire. As I got ready to go to bed, Ryan was on watch, waiting for the crew to pull up a first core from the Balearic Basin. Nothing eventful happened that night, and for the first time in the two weeks since we left Lisbon, I was able to get eight hours sleep.

Drilling went along rapidly at first through soft sediments. We took some spot cores on the way down and found the usual marine sedimentation. The first sign that we had hit something unusual came in the afternoon of the twenty-seventh. At the bottom of the sixth core we detected a few chips of anhydrite. Anhydrite, like gypsum, is a sulfate. It does not contain any crystallization water, however, and it has the composition $CaSO_4$. As mentioned earlier, laboratory studies have proven that calcium sulfate precipitated at higher temperatures tends to be anhydrite rather than the hydrous gypsum. The transition temperature depends upon the chemistry of the brine; it can be as high as 58°C in a pure $CaSO_4$ solution. The drilling rate slowed at this juncture, so we knew that we had gone through the top of the M-reflector, and that the reflecting layer here was anhydrite.

The next core we obtained was a laminated black mud, but the mud grains were so fine that we could not identify their mineralogy with the microscopes we had on board. Later Vlad Nesteroff determined by X-ray analysis that the sediment consisted of dolomite, one of the first minerals to precipitate after seawater has been somewhat concentrated by evaporation.

While Nesteroff and Wezel delighted themselves in photographic sessions with these exotic rocks, Cita and Stradner decided to get some rest. These muds, precipitated out of a sterile brine pool, contained neither foraminifera nor nannofossils to interest them.

The evaporite formation was hard, and drilling proceeded very slowly now. We were drilling at the rate of about a meter per hour. As the hours ground away, we began to wonder if we would ever reach basement. The drillers told us that we were not using the drill bit best suited for penetrating this type of hard rock, which made us think that perhaps we should pull out of the hole and drill somewhere else with a better bit. On the other hand we also began to realize the value of coring evaporites. We agreed to wait another twenty-four hours before making a final decision. Slowly and painstakingly the drill string continued to bore deeper. But frustrated by the slow progress, Ryan and I went to bed. It was about three o'clock in the morning of the twenty-eighth.

10

LOST SECRETS OF THE MEDITERRANEAN

—o—

WE WERE not to rest long before John Fiske, one of the marine technicians, roused us from our beds to admire the "pillar of Atlantis" (fig. 5). Ryan, Cita, and I shall perhaps always remember that day as the time we first saw the vision of a salt desert 3,000 meters below the sea level. We had enough facts to formulate a working hypothesis, but what we did not realize was that our findings were to explain many long-standing puzzles that H. G. Wells had referred to as "lost secrets of the Mediterranean" in one of his science fictions.

One of these puzzles was why six million years ago a biological revolution was sweeping across the Mediterranean Sea. The ancient marine fauna of the Mediterranean, descendents of mixed races from the Atlantic and Indian oceans, were effecting an unorganized mass exodus toward a refuge west of Gibraltar. Those that remained were soon to face extinction, except for a few hardy species that could tolerate the increasingly saline environment. Thus ended the Miocene Epoch, the less recent of the two dynasties that preceded the Quaternary Period. With the dawn of the Pliocene, or the more recent epoch, the emigrants returned and brought with them new species from the Atlantic. These were the ancestors of the present marine fauna of the Mediterranean. This dramatic event, as recorded by fossils in some sands and marls in Italy, did not escape the attention of Sir Charles Lyell, one of the founders of geology. The end of the revolution, signaled by the establishment of a new faunal dynasty, was designated by Lyell in 1833 as the historical datum dividing the Miocene and Pliocene epochs. What was the cause of this revolution? On that day in August we were too excited to recognize the possibility that the radical change in the salinity of the Mediterranean as it was being evaporated had induced the profound change in the marine fauna. Only later did we learn that French and Italian paleontologists had postulated a "Mediterranean salinity crisis" to explain the biological revolution. The idea had been quite popular during the early years of this century when H. G. Wells was studying geology with Professor Vincent Illing at Imperial College, London. Wells apparently did not weave his science fiction story out of thin air when he wrote that the Mediterranean was a deep hole in the ground "before the ocean waters broke in."

Also puzzling was a deep gorge buried under the plain of Valence in southern France that was discovered near the end of the last century during a search for ground water. The gorge was cut into hard granite to a depth of about a hundred meters below sea level. Filling the gorge were Pliocene marine sediments, which were in turn covered by the sands and gravels of the Rhone River. When the gorge was first explored, it was found to extend for some twenty-five kilometers between Lyon and Valence. Eventually this buried channel was traced more than 200 kilometers downstream to La Camargue in the Rhone Delta, where the rock bottom of the valley was reached by drilling to a depth of a thousand meters. It seemed that the Rhone River was merely a trickle compared with her giant and vigorous ancestor. What had induced the deep incision of the Rhone? On that day in August we were not even aware of the existence of such a question. Only a year later did we realize that a French geologist had anticipated our postulate. Denziot as early as 1950 had proposed the lowering of the sea level within the Mediterranean as the cause of the late Miocene erosion of the Rhone.

It would seem logical that the deep-sea drilling vessel *Glomar Challenger* had been sent to the Mediterranean to solve these mysteries. The fact, however, is otherwise. The Mediterranean cruise was designed to test other objectives. We just stumbled onto this surprising discovery.

The "pillar of Atlantis" was indeed hard rock. We recovered a few more "pillars" on the twenty-eighth, but the drilling rate was trying our patience. After pacing the rig floor all day, Ryan and I went to the mess hall for a midnight snack and then went to our cabin. But again we were not to be left undisturbed. At 0500 Anderson stuck his head in and told us quietly: "I think we got through the hard stuff!" We jumped out of bed and went to the core lab immediately. The technician on night watch had already taken the core in, and the sedimentologists had split it into halves, revealing some beautifully laminated sediments (figs. 6, *upper left*), which were indeed much softer than the anhydrite.

Stradner made a smear slide of the sample and examined it under the microscope. After half an hour he told us in his usual monotone that he had found abundant diatoms in the sediments. Diatoms are one-celled plants like nannoplankton, except that they have a silica (SiO_2) skeleton. Some live in oceans, but many other species are exclusively brackish or freshwater forms; they live or lived only in lagoons or inland lakes. Although Stradner was not a specialist, he had had enough general training in paleontology to tell us that the diatoms in our cores were not marine organisms. His preliminary finding was eventually confirmed by a specialist, Marta Hajos of the Hungarian Academy of Sciences, who

described both brackish and freshwater species from our Hole
124 samples.

How did these odd creatures get into the Mediterranean?
Ryan, not yet appreciating the meaning of the stromatolite
in the "pillar of Atlantis," was making a last ditch defense
of the idea of salt deposition from a deep brine pool. He
suggested that a brackish surface layer above a heavier brine
might have accommodated these diatoms. I myself was convinced, however, that a desiccated Mediterranean could have
been easily changed into a Caspian if there had been a sudden
large inflow of fresh water from somewhere. At the time I
was not sufficiently familiar with the European paleogeography during the late Miocene to know that there was indeed
a large brackish lake in eastern Europe then that could have
supplied a large amount of brackish water to the desiccated
Mediterranean.

Ryan's idea of a brackish surface layer eventually had to
be abandoned because Hajos identified in the cores from
Hole 124 not only diatoms that had floated in surface waters
but also several species that had lived on the bottom of a
brackish lake; these creatures proved that the brackish Mediterranean had been shallow enough to permit photosynthesis by plants growing on the sea floor. Thus we now have
no qualms about saying that the Balearic Sea at the time
when diatoms flourished was, like the Caspian Sea, brackish
from top to bottom. We had the last nail for the coffin when
we discovered the ostracod *Cyprideis* in diatomaceous sediments; none of these small mollusklike animals were ever
floaters; they could have lived only on the bottom of a brackish lake.

Anderson's optimism on the drilling rate proved prema-
ture. As the dawn came, the toolpusher told us that things
had gotten tough again; he was drilling at less than a meter
an hour. We were still debating the merits of quitting versus
continuing when we were relieved from making that painful

decision. Travis Rayborn came in and told us: "That's it. The bit is completely gone."

While on board ship we were much confused by the great variety of exotic rocks that had been hauled up: the "pillars of Atlantis," the stromatolites, the evenly laminated diatomaceous sediments, the marine marls. We did not have time to order our thoughts. Two years later, as I was revising the last draft of my report on the origin of the evaporites, I went to Lamont, where the DSDP cores were stored, and made a careful study of the undisturbed archive half from Hole 124. Only then did I discover several cycles of desiccation. The oldest sediment of each cycle was either deposited in a deep sea or in a great brackish lake. The fine sediments deposited on a quiet or deep bottom had perfectly even lamination (fig. 6, *upper left*). As the basin was drying up and the water depth decreased, lamination became more irregular on account of increasing wave agitation. Stromatolite was formed then, when the site of deposition fell within an intertidal zone (fig. 6, *upper right*). The intertidal flat was eventually exposed by the final desiccation, at which time anhydrite was precipitated by saline ground water underlying sabkhas (fig. 6, *lower left* and *lower right*). Suddenly seawater would spill over the Strait of Gibraltar, or there would be an unusual influx of brackish water from the eastern European lake. The Balearic would then again be under water. The chicken-wire anhydrite would thus be abruptly buried under the fine muds brought in by the next deluge. The cycle repeated itself at least eight or ten times during the million years that constituted the late Miocene Messinian stage.

But I am getting ahead of my story. The cores from Site 124 were to provide the key to our understanding many of these mysteries. On the morning of August 29, however, we were still undecided about whether to change to a more suitable drill bit and start a new hole here or to go to our next location. Ryan would have liked to have made a new start in the vicinity, but I was eager to move on to the eastern Mediterranean. I won the argument because we also had to think of a prearranged rendezvous with Bob Gilkey at high sea off Greece in early September.

11

INTERMEZZO

—⊖—

MARIA CITA's husband had taken two of their children sailing in the Aegean just before she joined us. The third stayed home with a broken leg. Ever since she had arrived on board, Cita had tried to find out if her sailors had returned safely. First she wanted to send a wire; but it was difficult to get the message out. Then we tried ham radio, but we could not make contact. When Bob Gilkey brought us Dumitrica, Cita repeatedly asked Gilkey to telephone her home when he went back on shore; he nevertheless did not appreciate the urgency of her request. Thus for weeks, Cita had been waiting in vain for a message from home. As we traveled to our next site, Cita's attention shifted away from the demands of daily routine, and she became more and more restless. Finally, I went to the captain, who thought we could telephone via Malta Radio.

On the morning of August 31, we sailed past Malta. Cita, Wezel, and Maync all got up at six o'clock to wait for the radio connection and by noon they had all spoken with their families. Cita talked to her oldest son, Nicolo, who assured her that he and Marco, her youngest, had had a good time sailing in Greece and all had returned safely. Pepe had had to change the gypsum cast for his leg. Everybody was happy and sound at home. Although none of us expected anything else, we all shared Cita's relief.

My wife, Christine, and I had hoped to communicate

through ham radio as we had on my earlier cruise. There had been no difficulty when I was in the South Atlantic, but to broadcast across the Alps was another matter. The radio operator and I tried to reach my friend in Bern every Saturday and Sunday at a prearranged time, but all our attempts had been futile during the last three weeks. Now the captain asked me if I would also like to call Zurich.

We called Malta Radio at 1130, 1330, and 1430 before an appointment was finally arranged for 2115. The appointed hour came, but nobody was at home when I called. It was a Monday night. Did Christine schedule her quartet evening at a friend's house? Did she go visiting? I became somewhat agitated. Malta Radio told me that we could try at 2215 again—which we did. When the call finally went through, I was relieved. Everyone was fine; Christine had just returned from taking Elee to a Mozart concert; Martin was doing better this semester in school; Andrew couldn't wait for vacation. Oh yes, Peter was growing; he could now reach the little bells that hung above the crib. Yes, she got my letters. She would come to Lisbon to meet the ship, and so forth.

Too excited to sleep after the news from home, I walked up to the electronic room. Our seismic profiler showed that we had crossed the Malta Escarpment. The sea bottom plunged several thousand meters from the Sicilian Channel to the abyss of the Sirte Basin (fig. 1). The M-reflector was with us all the way, and as usual, the reflecting surface ran parallel to the relief of the seabed underlying a thin blanket of ocean sediments that had been deposited during the last 5 million years. Ryan was right that the Mediterranean must have been a deep basin when the evaporites were deposited. Yet our cores told us that the brine pool had been shallow. How could we reconcile these two apparently contradictory sets of facts if this Miocene "death valley" had not been a hot desert 3,000 meters below today's sea level?

I wandered across to the bridge and chatted with the night watch in the dark. The traffic was light in this part of the Mediterranean and we saw no ship on the radar screen. Outside was a warm summer night. Sprawling on the deck I saw Orion shining overhead. I learned my stars when I was nine. It was the summer of 1939. After Chungking, the wartime capital of China, was all but destroyed by a fire bombing, our family sought refuge in a farm north of the city. For half a year, my sister and I did not go to school; we took our lessons from an elder sister. My father had a job then as an editor of children's textbooks. It was not his learned profession, but he gained access thereby to an excellent library. Every Saturday he rode a bus from the town to our village station and then walked five kilometers to the farm where we lived, lugging with him twenty or thirty books for us children. The temperature was usually in the nineties (above 30°C) but he always wore a heavy jacket to encourage perspiration to cool him off. Under the circumstances our consciences would bother us if we did not finish our weekly quota. Usually I read all that he brought. My favorites were books on exploration, particularly the accounts of Shackleton's ill-fated polar expedition, or the adventures of Sven Hedin, who walked in the deserts of Chinese Turkestan (Sinkiang) for seven nights, practically without water, before he finally stumbled onto a mudhole to rescue himself and his faithful camel driver. I wanted, then, to become an explorer. When I grew up I found that all the faraway places had been pretty much explored, and I was too old for the moon. Besides, I became accustomed to the comfort of home and turned into what my mother-in-law called in Swiss dialect, a *Stubenhocker*, or a "sitting-room squatter." It seems ironic indeed that I should finally fulfill my youthful ambition, exploring a new frontier, while being blessed with all the comforts of a luxury liner!

On the port side I could see the Milky Way. In Chinese it

had the more romantic name Silver Stream and a legend of a tragic love. On one bank of the stream was the virgin maiden. Punished by her aristocratic God-Mother for having fallen in love with a cowhand, she was confined to a weaving stool all year long. But on the evening of the seventh day of the seventh month, the crows would come and build a bridge so that she could meet her lover, who had to stay where he belonged on the other bank. This August night I searched in vain for the weaving maiden (the star Lyra) and her lover (Aquila). But they were gone, for it was already four in the morning. Realizing the lateness of the hour, I got up and went back to my cabin.

12

THE DELUGE

—o—

Glomar Challenger was not a fast ship. She claimed a maximum cruising speed of 12 knots, but we were moping along at about 10 knots. It took almost three days to reach our next drill site, but this breathing spell was put to good use. Shipboard reports for each site had to be written, typed, edited, revised, and retyped. Staff conferences were called to review our findings and to make future plans, now that we had reached the halfway point in our cruise.

Cita, Wezel, and Maync had all been taught more than the rudiments of Mediterranean geology. They were all familiar with the various occurrences of evaporites on the land surrounding this inland sea—the late Miocene evaporites in Spain, in Piedmont, in Tuscany, in Marche-Umbria, in Calabria, in Sicily, on Ionian Islands, on Crete, on Cyprus, in Israel, in Algeria, and so on. Like others in our profession, however, they had considered these local lagoonal deposits, formed during an epoch of unusually dry climate. As we now steamed eastward, relieved from the tension of daily drilling operations, the significance of our discovery began to sink in: we had found an evaporite formation beneath the Mediterranean sea floor, and we had geophysical evidence to suggest that the whole Mediterranean was underlain by this formation. With this knowledge we could no longer dismiss the late Miocene evaporites on land as simply local deposits.

Both Cita and Wezel had studied the evaporite formation

in Sicily, the *solfifera sicilienne*, and knew it to include a thick series of rock salts, gypsum, and anhydrite. Interbedded with the evaporites were marls with Messinian fossils. The youngest evaporite formation from other circum-Mediterranean countries were also Messinian in age. Unfortunately, the fossils in our cores from Hole 124 were not very good, and Cita had not yet had a chance to take a close look at them. She was eager to get some better samples at our next site to confirm the Messinian age of the evaporite deposition in the Mediterranean.

The *solfifera* in Sicily is overlain by a white oceanic sediment called *Trubi* marl. This marl has a microfauna that could have lived only in a deep sea of normal salinity. Paleontologists are quite certain of their conclusions: the *Trubi* foraminifera belong either to species that swam in deep waters or genera that dwelled on the deep and cold sea bottom. Later we were told by Dick Benson of the Smithsonian Museum, a specialist in ostracods, that the *Trubi* marine ostracod fauna in this formation is also a typically cold water assemblage, similar to those forms that live in the deep Atlantic today. Yet as the editor of *Sedimentology*, I had just processed, prior to my departure for Lisbon, a manuscript by Laurie Hardie and Hans Eugster of the Johns Hopkins University on the *solfifera sicilienne*, which claimed that the Sicilian salts were laid down on a playa. If both the sedimentologists and the paleontologists were correct, the inescapable conclusion would be that the Messinian salt deposition in Sicily had been ended by a sudden deluge. When marine waters returned, the salt pan was turned suddenly into a deep sea. At the previous site we had failed to obtain a core that recorded the passage from evaporite to normal marine sedimentation. Nevertheless our oldest Pliocene core there showed considerable similarity to the *Trubi* marl of Sicily. Was it a clue that the deluge was not a local event in Sicily? Perhaps the whole Mediterranean was drowned when the floodgate at Gibraltar was crushed. The evidence was tantalizing, but we had to be sure. There was no alternative but to sample meter by meter at our next site with continuous coring.

After three days of steaming across the Mediterranean, we finally dropped the buoy that marked Site 125 to the Ionian sea bottom, about 300 kilometers southwest of Crete (fig. 2). For the first time in our cruise, we managed to arrive on station without a crisis. The location helped. We were to drill anywhere on the flat-topped submarine ridge—the Mediterranean Ridge. It did not matter if we missed our target by a kilometer, or even ten kilometers. Not having to work under pressure, everything functioned perfectly.

We chose a site of slow sedimentation for continuous coring. Here on top of the ridge, skeletons of nannoplankton and foraminifera accumulated at the rate of some two centimeters per thousand years. We could thus obtain a record of 5 million years of geologic history by taking one hundred meters of cores. To have a continuous history, however, we had to have 100% recovery of coring, and it was our misfortune to run into trouble again here.

The drill string reached the sea bottom shortly after midnight on September 1, and the initial operations proceeded rapidly and smoothly. We recovered four full barrels of Pleistocene (very recent) sediments. All of us were excited by their beautiful coloration: alternations of green, orange-red, brown, and gray-black. Cores were coming up at a furious pace, but our civilized French colleagues in sedimentology were not disturbed by the rush. They had their two-hour lunch break and enjoyed their leisurely steak dinner, while the backlog of unprocessed cores piled up.

Trouble was apparent around 0915 after the fifth core arrived on deck with only 30% recovery; the other 70% of the barrel was "empty." Anderson thought the poor recovery had been caused by an abnormally short plastic liner. So a longer liner was put in. The situation did seem to improve, as the sixth and seventh cores had somewhat better recovery. But we knew something was definitely wrong when we found the eighth core almost completely empty. I talked to Jim, the toolpusher on duty. He thought that perhaps not enough weight had been put on the drill bit. A change was made accordingly when Ryan came to relieve me.

I got up a few hours later to find that things had gone horribly wrong and that we might have to pull out of the hole. As Ryan told it, the ninth core was also a bust. Nothing came up except a few lumps of ooze in the core catcher. This time, Charlie, the alternate toolpusher, ventured a diagnosis that eventually turned out to be correct. Anderson had decided to put in a "flapper valve" at the bottom of the drill string as a precautionary measure against the intrusion of loose material into the drill collar, a problem that had plagued us in the previous two holes. Such valves had been used from time to time with mixed success, and we were experimenting with a new modification. If the valve had functioned properly, it would have sprung open when the core barrel was down so that sediment could find its way into the barrel. After the full barrel was taken out by the sandline, the valve would have flipped into a closed position to prevent sands and other loose material from entering the drill pipe. If the flapper valve malfunctioned, however, the core barrel would be shut all the time. No sample could get inside, and we would find an empty barrel when it was hauled on deck.

In fact we did not expect to encounter (and did not) any sandy formation at this site, so no flapper valve was needed. After listening to Charlie's diagnosis, Ryan and I were ready to accept his advice that we waste no more time; we should pull up the drill string immediately and remove the defective flapper valve. Anderson, however, overcautious as ever, wanted to have another try. So a tenth core was attempted, but nothing was recovered.

In early afternoon Ryan and I again asked Anderson to raise the drill string. He was still not convinced. Perhaps the fault lay with a defective core catcher, he thought. He was grasping for straws. Anyway, another barrel was sent down with a brand new catcher. Exasperated, Ryan decided to attack the backlog of cores piled up in the core lab while the sedimentologists dined. One after another plastic liner was split open. Finally, Ryan located the culprit: in one of the cores he found a piece of a broken spring from the flapper valve. Just then the crew hauled up the eleventh core. It was again empty, but nobody was surprised. At 1900 hours, some ten hours after the first signs of trouble, Anderson finally gave orders to bring up the drill string. The operation of dissembling and reassembling the string continued for another twelve hours, and we did not resume coring until 0900, September 3. It was a cheap valve, but an expensive failure; we had lost $50,000 worth of ship-time.

So we started all over again in the same place on the third. We had greater success this time, but our recovery was still far from satisfactory. Then, after drilling through some eighty meters of soft oozes, we hit the evaporite layer, and the drilling rate slowed to almost a halt. By 2000 hours we were progressing at only about a meter an hour. Worse still, when the next core came up, the barrel was again empty. Our toolpusher thought that this time we had been too heavy-handed with the pump pressure; core materials had been washed away by the circulating fluids. So we shut down the pump for coring, but that led only to a jammed drill pipe. Mud was pumped down and finally the drill pipe was pulled free, but the core barrel was stuck inside the pipe. We sent down a sandline to fish it out. It was no use. The shear pin broke and the barrel remained firmly set. Again we were at the end of the road. We had to pull out of the hole. Later we were reproached by many of our colleagues on shore for not having tried to drill deeper at this site and others. Like other "Monday morning quarterbacks," however, they had too little appreciation of the operating conditions. In fact, I myself began to forget all our woes after I had been back ashore a couple of months; only my operations diary served as a reminder.

After the cores were investigated and the data processed, it turned out that we did not fare too badly at Site 125. Combining the cores from the two boreholes at this site, we managed to put together a continuous section for the sedimentary record of the last 5 million years, just as we had hoped. And more important, we obtained a core that recorded the end of the evaporite deposition and the beginning of the deluge. The last sediments of the Miocene were carbonate mud, which contained only very poorly developed forms of marine organisms. We now interpret this mud as the sediment laid down during the transient stage when the desiccated Mediterranean was again filled up. The salinity of this rising sea must have been higher than normal, for unlike the present situation, where heavy brines produced in the Mediterranean by excess evaporation can find their way back to the Atlantic through the Strait of Gibraltar, the latest Miocene Gibraltar was the site of a huge waterfall, a one-way street. There was no way for the heavy brines to flow back into the Atlantic. The salt water grew more and more salty under the hot and sunny Mediterranean sky so that only some dwarf microfaunas could survive. Those were the last of the Miocene Mediterranean creatures.

Suddenly, at the end of the Miocene, the dam broke and the basin was quickly filled to the brim. Then the influx of the Atlantic water and the reflux (return-flow) of the Mediterranean brines were able to moderate the high Mediterranean salinity caused by evaporation. The Mediterranean became again a habitable environment for marine organisms. The Pliocene immigrants from the Atlantic constituted the *Trubi* faunas, and they were mostly new species. Only a few could trace their ancestry back to the creatures that had lived east of Gibraltar prior to the catastrophic desiccation. Those organisms returning were descended from the Miocene refugees to the Atlantic, where the lineage survived. The offspring of the Pliocene newcomers never had to face a salinity crisis, and many of them are flourishing in the modern Mediterranean. It is thus not surprising that the present marine organisms of the Mediterranean bear great resemblance to Pliocene fossils and are quite distinct from Miocene fauna, which largely perished during the Messinian salinity crisis. This finally provides the rational basis for Lyell's distinction of the more (*Plio-*) from the less (*Mio-*) recent faunas.

13

IN THE BEGINNING THERE
WAS AN OCEAN

—o—

IT WAS a tedious job to pull up the drill string by disconnecting one pipe segment from another, thus it was not until almost midnight of September 3 that they finally brought up the drill collar. The core barrel was jammed inside by stinking debris. Training a hose on the drill collar, the roughnecks washed away the mud and guck before pulling the core barrel free. A helpful marine technician scooped up some of the "dirt" from the rig floor, and that turned out to be the only sample we had of the evaporites from the eastern Mediterranean.

We got stuck at Site 125 under precisely the same circumstances we had encountered at Site 122. We drilled into the top of the evaporite formation and ran into gypsum, using the wrong drill bit. Instead of a "button bit," which broke a formation apart by concussion, we were using a "tooth bit," which was effective only in digging soft oozes. So we could not penetrate the evaporites effectively. The little headway we did make served only to chip off fragments from gypsum crystals, and these chips had been the source of trouble at both sites. If we used a strong pump to wash them away, we would have nothing left in the core barrel. But if we shut off the pump as we had, the fragments would be mixed with mud and would get inside the drill collar and jam the core barrel. Previous deep-sea drilling cruises had never encountered any of the evaporite formation, thus nobody had ever had any experience with this sort of problem. By the end of our cruise, we were experts, but it was too late to make much difference.

A sulfurous smell permeated the core lab with the recovery of the jammed core barrel. Apparently the gypsum had been changed into sulfur in part by sulfate-reducing bacteria. On-shore chemical investigations later confirmed our nasal diagnosis. In fact we were then told that the sulfur deposits in the *solfifera sicilienne* also owed their origin to this sort of bacterial action. We had apparently drilled into the underwater equivalent of the *solfifera*.

While we were having our midnight snack, Ryan and I talked over the drilling situation. We had hoped to decipher the whole history of the desiccation. We were convinced that evaporite sedimentation had terminated at the end of the Miocene, some five to five and a half million years ago. But when did it begin? Our last four attempts to find this answer had all been frustrated. Nature did not yield its secret lightly. Since we could not get through by hitting straight down the middle, we decided to try an "end run."

There is a deep cleft in the Mediterranean Ridge some 120 kilometers northeast of Site 125, which was discovered on the seismic record made by the research vessel *Conrad* of Lamont in the late sixties (fig. 31). It is not a fracture, nor a fault, for the beds on the two sides of the gorge have not been relatively displaced. It resembles, rather, an erosional feature. The record showed that erosion had not only removed soft sediments but had cut through the M-reflector as well. Ryan had sailed on that *Conrad* cruise. He and his colleagues could hardly believe that the cleft was erosional. They knew of submarine canyons sculpted by underwater currents, of course, but this one cut across the Mediterranean

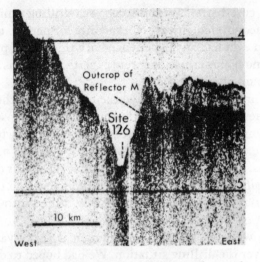

31. Seismic record from the research vessel *Conrad* of Lamont, showing the deep cleft in the Mediterranean Ridge cut by a river of salt water rushing in from the Atlantic to fill the desiccated Mediterranean basins.

Ridge. What kind of erosive agent at the bottom of the Mediterranean could be so powerful as to breach the M-reflector?

We had learned to respect the M-layer. This evaporite formation had resisted all our attempts to penetrate its depths by drilling. Submarine currents drifting across the ridge could never have been strong enough to cut through such a tough formation. As our story of a desiccated Mediterranean began to take shape, however, we could imagine a great river carving this deep gorge. We could picture a time in the late Miocene when the Ionian Basin was being filled up to the level of the ridge crest, which separated the basin from the deep Hellenic Trench to the north. The seawater would first spill over a saddle on the ridge and then cut into the hard rock to make a channel; this would ultimately reach a dimension comparable to the Yangtze Gorge of China. Erosion would stop after the Mediterranean was drowned, and indeed the gorge is now buried under one hundred meters of soft sediment.

This buried wonder of nature gave us the chance for an "end run." Instead of fighting the M-layer here at Site 125, we could relocate over the cleft, where the hard evaporites had been removed by erosion. There we should be easily able to drill through the soft ooze cover to sample the old sediments beneath the evaporites on the wall of the gorge.

Our decision to move to this nearby location was made just before dawn, thus we had only a few hours of sleep before the captain sent an able-bodied seaman to awaken us. We were to drill into the side of the buried gorge at this site, where the soft sediments were thick enough to stabilize the drill string, but not too thick to hinder penetration. So precise navigation was again required. It was thus not surprising that our sat-nav set quit on us again and induced another heated confrontation between the co-chief scientists and their technicians when the chance for misunderstanding presented itself. We seemed jinxed by our unlucky number. And wasn't today Friday, September 4? We all know that four and nine make another thirteen!

We finally dropped the beacon at Site 126, and drilling started around 1600. By 2215 we had already penetrated one hundred meters and recovered three barrels of Pleistocene sediments. But at this point the trouble started again, for we began to hit the hard stuff. The drill string simply could not find its way down. Instead of going to bed, Stradner, Cita, Ryan, and I all waited for that core, which we expected to tell us the story of the Mediterranean before the crisis struck. We were finally to get the answer we had been seeking the last two weeks. We were impatient indeed and tried to while away the time playing ping-pong. Then it got too hot in the gym, so I quit and went to cool off at the bow.

The new moon was up and cool breezes blew in from the northwest. I looked up at the tower, where the drill stem turned around and around, but it seemed to have made no progress downward. Returning to the driller's shack, I was told that we had only cored three meters in three hours. We decided not to wait until the full nine-meter barrel was cut. The sandline was sent down and the core barrel fished out after midnight. Imagine our disappointment: the core barrel was empty. Empty! Not even a lump of ooze, or a speck of dirt! Completely frustrated, Ryan and I went to our cabin, leaving an order to drill ahead.

I was half-awake at eight o'clock but was unwilling to get up to face new disappointments. Ryan must have felt the same. So for the first time in more than a week we had eight hours of sleep. Finally, Anderson came in at half past ten. As we had expected, progress had been extremely slow. Anderson thought we had hit anhydrite again. He felt that "no recovery" and "slow penetration" were sufficiently diagnostic signs and proposed that we pull up the string to change the drill bit if we again received no recovery from the next barrel. I thought we were drilling into a considerably deeper horizon than anhydrite, but maybe digging in some hard rock like chert. Ryan thought we should move the location to a point where the cleft was deeper in order to reach older rocks beneath the valley fill. As we argued, the next core was being hauled up. When it finally reached the deck at about 1100, we ran to the rig and saw water spilled out of the core barrel. I caught a few rock chips with my bare hands and ran down to Stradner in the paleo lab. Actually the recovery was not bad. We got enough material to find out that the sediment beneath the evaporite was a dark marl, a normal marine pelagic sediment of middle Miocene age (some 12 to 14 million years old). So the Mediterranean, or at least its eastern part, had been a normal inland sea before its outlet or outlets were completely shut off in the late Miocene. What was it

20 million years ago? What was it 100 million years ago? When was its beginning? For the time being we had to suppress temporarily our excitement over this discovery and discuss the more practical question of getting maximum results by drilling.

We did some calculations. Normal accumulation rates of marine pelagic sediments should be on the order of two centimeters per thousand years. We could read back 1 million years of history through a penetration of twenty meters and 5 million years through one hundred meters. If we continued at the present drilling rate of two meters or less per hour, it would cost us two whole days of ship-time to look back another 5 million years. It would not be worthwhile to proceed at this rate, particularly because we had reason to believe that no especially dramatic event had occurred during that 5-million-year interval. We wanted very much, of course, to get a glimpse of the beginning of the eastern Mediterranean, but we could never penetrate deep enough here to reach that objective.

The fact of the matter was, as we were to learn later, that we had again used the wrong drill bit. We had changed to a button bit when we started drilling Hole 126, when we should have used a tooth bit to tear apart the soft waxy shales. In such a situation we should have raised the drill string, as Anderson advised, and changed the bit, but we hated to invest another day of ship-time for an uncertain venture. After much debate we finally took up Ryan's suggestion to move to another site where the gorge might be cut deeper.

The decision was communicated to the captain, and the vessel moved to the new target. Drilling resumed in the afternoon. Our first surprise came when we hit the hard layer at 60 meters below the sea floor, not at 200 meters as we had predicted. Obviously the gorge had not cut deeper here. All evening long the driller ground away at a maddeningly slow pace, while Ryan and I took turns "sitting" on the rig floor. For the first time during the cruise we felt so frustrated that we became somewhat uncertain what our next course of action should be.

The first core from the new hole came up at 1900. We were sampling the same dark marl. In fact we were actually coring a younger middle Miocene strata. Ryan and I then decided that there was no use spinning our wheels in this waxy shale. We asked Anderson to give orders to pull out of the hole. Then we went to the ship's office to see the captain.

The captain had planned to effect a rendezvous with Bob Gilkey at this site, and a confirming message from Gilkey had just come. Yet the captain understood our predicament; we could not just sit here until Gilkey arrived. The rendezvous had to be made, of course, but we could send a message to direct the U.S.S. *Buttes* to our next site.

Before we had completed our arrangements with the captain, the three-member paleontology staff stormed into the ship's office. They were indignant to have heard from the roughnecks that we were abandoning the hole. "Why should we leave now that we finally found a place where we can drill into the rocks beneath the evaporites?" Maync demanded.

We gathered in the science lounge for a conference. Yes, they were right, we should know more about those rocks. Yes, we wanted to stay also, but we just did not seem to be able to get any deeper. Yes, we knew that the rock was soft and waxy and that you could scratch it with your fingernail, but we did not seem to get any traction; we were spinning the drill stem without getting anywhere. We asked them not to be discouraged. We had recovered a sample that proved that there had once been an ocean here. Other samples told the story of the gradual isolation of the Mediterranean Sea as Africa collided with Asia some 20 million years ago, cutting off the connection between the Mediterranean and the

Indian Ocean. We could now refer to similar rock sequences on land in western Greece to deduce the earlier history of the eastern Mediterranean. Finally, we promised them that we would take another crack at the M-reflector north of the Nile Delta. We hoped we would get a better section there. Our colleagues were gradually reassured. To bring the evening to a happy ending Cita donated a bottle of port from Portugal.

14

A MEDITERRANEAN
SEABED TUCKED UNDER
AN ISLAND

—O—

THAT we had found evaporites beneath the Mediterranean seabed at a depth of thousands of meters below sea level was in itself no proof that the floor of the salt basin had always been that deep. Vlad Nesteroff favored an alternative hypothesis. He believed that 5 or 6 million years ago the Mediterranean had been a shallow shelf sea like the Baltic today, which then became a shallow salt pan when its connection to the Atlantic was severed. According to Nesteroff's postulate, which found considerable popularity among his French colleagues, the Mediterranean had somehow foundered during the last 5 million years.

Nesteroff expressed this idea in his shipboard report, and he wrote a separate chapter on the origin of the Mediterranean evaporites in our final cruise report. He did not really have any evidence to support his claim that the Mediterranean had been shallow; but he thought the idea of a shallow salt pan was more credible than a deep desert basin. Facts, however, contradict the idea of a shallow Mediterranean during the late Miocene; they speak unmistakably in favor of the more surprising idea.

The first and most obvious support for the concept of a deep Mediterranean basin came from a study of the seismic records. The M-reflector had been discovered before the Leg 13 expedition, and everybody was then convinced that the sediment constituting the reflecting layer had been laid down in a Mediterranean basin whose topography was not much different from the bathymetry of the Mediterranean today. Except for some local disturbances the Mediterranean seabed 6 million years ago lay at about the same depth that it does now. In fact this was Ryan's reason for having once argued for a deep water origin of the evaporites. Other evidence was provided by Cita and the other shipboard paleontologists. The fossils in the sediments immediately underlying, immediately overlying, and interbedded with the evaporite beds all represented deep water creatures.

One final reason for our not accepting the shallow bottom hypothesis derived from our knowledge of the geological history of the Mediterranean. We had good reason to agree with Argand that the Balearic Basin had been created some 25 or 30 million years ago, long before the salt had been laid down. Furthermore, our geophysical studies had indicated that the eastern Mediterranean was even older and might date back to the Mesozoic Era, some 200 million years ago. During the last 5 million years, the eastern Mediterranean had not foundered, which would have required regional tension. On the contrary, the sea bottom had apparently risen under compression as Africa and the eastern Mediterranean seabed were pushed northward toward Europe. Our next drill sites in the eastern Mediterranean were primarily designed to test this hypothesis of compression.

Site 127 was targeted to investigate the northern edge of the Hellenic Trench (fig. 2). Trenches, which are linear depressions in the ocean floor, are commonly found on the edge of continents or on the ocean side of island arcs. The Peloponnesian peninsula and the islands of Crete and Rhodes constitute the Cretan Arc north of the Hellenic Trench. According to plate tectonics, a trench is present where an ocean and an island arc meet under compression. The ocean floor is pushed down to form the trench and the island arc is thrust up (fig. 18). The intermittent movements of the sea floor plunging beneath the arc are sufficiently powerful to cause earthquakes. From the study of these earthquake movements one of our JOIDES friends, Dan MacKenzie of Cambridge, had suggested that the Mediterranean is now moving northward under the Cretan Arc. If the idea were correct, we should find ocean sediments under the rocks that form the foundation for the island of Crete.

The idea that the Cretan Arc had been thrust up can be found in Greek legend. It was said that Apollo was short-changed when the gods of Olympus divided Greece. As an inexpensive gesture of generosity and consolation, Zeus promised Apollo the ownership of the islands that would rise from the sea. One fine day as Apollo was flying over the Aegean, he saw the island of Rhodes emerge from beneath the sea. He descended to claim the island, and the famous Apollonian Temple of Lindos is built on the spot where he landed.

There is very good scientific evidence to suggest that the island of Rhodes did in fact rise from the sea. The Lindos temple was built on a newly emerged marine terrace. The island of Crete also seems to have risen from the sea. On that island we found late Miocene evaporites and marine sediments which had been deposited on a submerged seabed before it was thrust above sea level sometime in the Pliocene, about 3 million years ago.

Glomar Challenger departed for the Hellenic Trench site at 0500 on September 6. I felt the vibration as the thrusters started again. Shortly before ten o'clock the chief mate stuck his head in and told us that we were approaching our target.

It had been another short cruise. Ryan had difficulties opening his eyes after having stayed up all night talking to Anderson about "hippies." It bothered some people on board that all the marine technicians dressed and had long hair like hippies. Yet we all found them responsible, hard-working, and likable young men. They did present an interesting contrast to the drilling crew: no one was likely to mistake a long-haired Scripps technician for a hard-hitting roughneck from Louisiana or Mississippi. Despite this incongruous mixture no serious conflict ever broke out during all the DSDP cruises. Long sea voyages must have taught us patience and tolerance.

I went up to the electronics lab at ten o'clock sharp and found the vessel steaming northeast toward the waters above the Hellenic Trench. The PDR showed a sharp descent on the bottom profile from the ridge to the trench (fig. 32). We had secured good sat-fixes all morning, and we had kept close track of our position by dead reckoning, for precise locationing was absolutely necessary here to insure success. Our goal was a spot at the foot of the northern trench wall. We had decided to cross the floor of the trench, release the buoy, secure our gear, and then return to the targeted site.

Shortly after 1100, we were sailing over the flat trench floor. The captain dropped in every few minutes to give us the countdown: "We are about ten minutes from the target by dead reckoning," he reported. "Eight minutes. . . ." "Five minutes. . . ." "Three minutes. . . ." Ryan, Pautot, and I watched the PDR record intently, waiting for the first appearance of a side echo, which should show up when we got close enough to the trench wall. Pautot was to give final signal to drop the buoy.

"Now?" Ryan asked prematurely.

"*Non.*"

Tick, tick, tick—half a minute was gone. We had gone 150 meters farther.

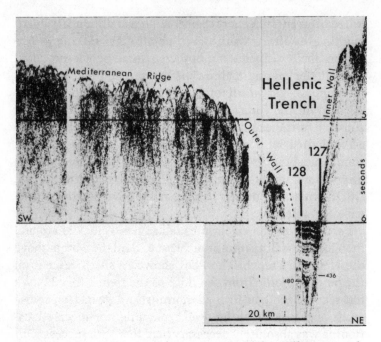

32. Seismic record showing a cross section of the Hellenic Trench, which runs between the island of Crete and the Mediterranean Ridge and represents the deepest portion of the Mediterranean.

"Now?"

"*Non.*"

Tick, tick. Another echo, and another 150 meters.

Finally we saw parabola-shaped figures rising above the flat profile of the trench floor; they were the first side echoes. The trench wall was so steep and we were so close to it that the echoes that appeared on our register included not only those reflected back from the flat trench floor directly below the ship but echoes from the steep wall adjacent. The appearance of the side echoes told us that we were getting very near our goal.

Tick, tick. The echo ticked away, when Pautot finally shouted, "*Mais oui!*"

The signal was transmitted over the phone and the buoy was dropped. When the next tick came, the bottom profile rose sharply; the first bottom echo had bounced off the foot of the steep trench wall now directly below us. Ryan was pleased that we had navigated to within 200 meters of our target. Some maneuvering continued before our position was fixed by the beacon at 1200. Later our success in positioning the ship over this site was dismissed as "incredible" luck; the "Monday morning quarterbacks" gave no credit to our talented navigators.

The drill string touched down in the evening of September 6 at 4,664 meters below the rig floor. I had the night watch. For two nights and a day cores were hauled on deck, consisting of barrel after barrel of sands and muds. There was a flurry of excitement in the early morning of the eighth, when the U.S.S. *Buttes* brought Bob Gilkey, supplies, and staff replacements—a maneuver code named "Operation Broadhaul," since the two technicians who came to join us were both women. Mail also came, but I was disappointed to find that letters from home apparently had not reached Scripps in time to be forwarded to me.

Shortly after ten o'clock in the morning of the eighth, I left instructions with the toolpusher to drill ahead and went to bed, having stayed up all night. But something exciting always happened when I went to sleep. After two hours Ryan came in and switched on the light. He had a piece of rock in his hand and was extremely agitated: "Ken, we hit pay dirt!"

The rock was a gray, well-lithified limestone. It appeared to be much older than anything we had encountered in the Mediterranean so far. Some thought it was part of the evaporite sequence. Others, including Ryan and me, thought that it was much older, probably Mesozoic, or some 100 to 200 million years old. Maync cut slices of the limestone and made thin sections, then we all looked at them through the microscope and searched for signs of fossilized animal bodies. We did not search in vain. In the early evening, I found a foraminifer and we called Cita.

The organism belonged to the genus *Ammobaculites*, a genus that had lived from the Carboniferous Period—300 million years ago—to the Recent. It was thus useless to resolve our dispute. Nevertheless this discovery did encourage Maync to cut more rock slices, make more thin sections, and look at them more thoroughly through the microscope. The effort paid off, for he reported the next day that the limestone was indeed Mesozoic. It contained *Orbitolina*, a group of foraminifera that had lived in the early Cretaceous Period, some 100 to 130 million years ago. There was no more doubt that we had gone through the young sedimentary cover at the bottom of the trench and touched the rocks underlying the Cretan Arc.

We really hit pay dirt with our next core, which sampled the Pliocene Mediterranean seabed under the foundations of Crete. It was exactly as the theory had predicted. The Mediterranean floor had been thrust under the Cretan Arc during the Pliocene, as Africa was pressed closer and closer to Europe. This proved that the eastern Mediterranean had been under compression during the last few million years and discredited the hypothesis that the region had foundered under tension. Our next goal was to see if this compression had also been responsible for the birth of the submarine mountain called the Mediterranean Ridge.

15

MÉLANGE UNDER STRABO MOUNTAIN

——O——

FOR SEVERAL days, during which we had drilled another site in the Hellenic Trench, I had been nursing a mouth sore, and now I was getting a bad toothache. Since we were on our way to the next site, I could go to bed—which I did, thinking I would feel better after a good night's sleep, a rarity since we had left Lisbon. But the toothache did not go away, so I went to see the ship's surgeon the next morning complaining of my bad tooth. He treated the sore and declared that everything would be all right in a few days.

We approached Site 129 early in the afternoon of the thirteenth. It was another trench site, located at the foot of Strabo Mountain, a submarine peak. Again we had to locate the ship precisely over our target, but again strong currents hindered our progress. We had to continue our maneuvering until dusk before our position was finally stabilized. I got up early on the fourteenth, feeling much better, had a hearty breakfast, and relieved Ryan on watch. Drilling operations proceeded smoothly. We cored a middle Miocene marl dating some 25 million years. There had to have been some earthshaking process to lift Strabo Mountain up so that such old rocks were found close to the sea bottom mixed with younger ones. To make sure that this rock was not just a loose block fallen down from the steep wall of the "mountain," we spent hours drilling another fifty meters and took another core. We got the same black marl. So we were indeed drilling through the bedrock. Our finding here confirmed the discovery made at Site 126: the Mediterranean had been a broad seaway connecting the Atlantic and the Indian oceans before the late Miocene salinity crisis (fig. 25).

At this stage Ryan and I agreed that we should explore other parts of the wall by drilling one or more offset holes. Meanwhile my toothache grew worse, so I went to see the surgeon again. He gave me penicillin to prevent possible secondary infections but insisted that it was the mouth sore and not a tooth infection that was bothering me.

I retired early to try to get some sleep, but the pain kept me awake. Restless, I got up several times during the night to check with Ryan and Cita on the progress of drilling. The results were interesting indeed. Ryan had moved the *Challenger* and had drilled an offset hole on the northern slope of Strabo Mountain. Our first core there gave us a piece of green marl, which contained numerous ostracod fossils. These ostracods were tiny mollusklike creatures, and the species we found had lived only in brackish waters. According to our seismic profiling record, this marl unit underlies the M-reflector and belongs to the evaporite formation. The finding in this offset hole was in line with our discovery at Site 124: the desiccated Mediterranean basins were turned into brackish lakes when freshwater influx exceeded evaporation.

Even the excitement of this discovery did not dull the pain of my toothache. Sitting in Cita's lab at four o'clock in the morning, Ryan and Anderson discussed the possibility of getting me to a dentist. We were only about 80 kilometers off Crete. I demurred, not knowing how long we would stay at this drill site; they might have to wait for my return. After I had yelled all during the trip that no time should be wasted, I would be the last to permit a loss of ship-time on my account. Were there other possibilities? I was advised to consult the captain. I went up to the bridge shortly after five but the captain would not be up until six. I did not want to disturb him. There had been talk of the possibility of running into an American naval vessel with a dentist on board. But the third mate made clear the improbability of such an event. Exhausted and despairing, I went to bed and actually slept. An hour later the captain woke me up. He had talked with Ryan and with the surgeon. He told me that they would have taken me to Crete in the *Blue Fox* if I really had a bum tooth but that the surgeon had insisted that I was bothered only by a mouth ulcer. I knew I had a tooth infection, but I did not argue, because by now I was fairly certain that a trip to Crete would affect our drilling plans; I did not want to mess up the program. Anyway I went down to see the surgeon again, and he took another look. The sore was now almost healed. Finally he admitted the possibility of an infection under the filling of my lower left molar, where I had felt acute pain for several days. I was given all kinds of medicine—antibiotics, morphine, tranquilizers. I swallowed one of each and was able to doze off all day.

With my being laid off, Ryan had to take over the planning and supervision of our operations almost completely. I went to the core lab occasionally and joined discussions, managing to give Ryan moral support from time to time. I had not realized before I assumed this tour of duty that co-chief scientists would be called upon to make so many decisions. Some were minor; others were critical. Before drilling we had to recommend the drill bit to be used, a decision that often determined the success or failure of a hole. And we had made the wrong decision more than once! We also had to define the coring schedule—whether to take surface cores; go down twenty, or seventy, meters for coring; core continuously; core back to back and then skip some intervals. Should we drill with the core barrel inside the drill string? Should we take the barrel out and send in the center bit?

Should we cut a full barrel, or should we raise the barrel after two or three meters? Ryan and I complemented each other very well: I tended to emphasize drilling deeper, whereas Ryan was more concerned that we not miss critical samples. Yet we always managed to compromise.

One of the hardest decisions to make was when to quit. Eventually, as we got to know the capability of our drilling equipment, both of us would come to almost simultaneous conclusions when the drilling became too slow to be productive. This experience was not shared by all members of the scientific staff, however, and sometimes, when we had to make quick decisions in order to prevent further waste of time, our reasons were not explained to our colleagues. This failure engendered the protest led by Cita when we abandoned Hole 126. Our colleagues normally gave us their full sympathy after we had explained a decision. Yet we did not always have time to explain everything. So some days our colleagues might wake up and find themselves at another site drilling a new hole. Not knowing the reason, this could be irritating.

Before we started drilling at Strabo Mountain, Ryan had discussed the strategy with me. We knew we would drill into compacted sediments or even hard rocks. We also expected to find various formations chaotically placed, piled one on top of another by compressional forces to form a mélange of different rock types. Since we could not hope to penetrate within a day or two more than a hundred meters into a hard formation, we decided that we could gather the maximum amount of information by changing drilling locations to sample the top of the randomly distributed rocks of various kinds and different ages. Making an offset cost little time as long as we did not have to pull the whole drill string back on deck. And it would be more profitable than trying to bull our way through a massive formation.

Our first hole at Site 129 penetrated into the type of rock that had stopped our drilling at Site 126. We had used a different type of drill bit this time, which should have yielded better results, but we reached almost a complete halt after drilling through only one hundred meters. Ryan decided to drill an offset hole, but he met strong opposition. I got up just in time to join the debate and give him the support he needed to proceed. Eventually we drilled three offset holes at the Strabo Mountain site and made a different discovery at each. If we had continued to drill at the original hole, we would probably have drilled deeper into the same formation and missed all the other exciting information. This was just another illustration of the decisions that had to be made; we were perhaps lucky that we did not always choose the wrong one.

16

YESTERDAY A MARSH,
TOMORROW A MOUNTAIN

—O—

THE TITLE of this chapter was the headline in *Figaro*'s report of our cruise. The "marsh" referred to the sabkha on which the Mediterranean evaporites had accumulated. The "mountain" was not yet born but was rather being developed in embryo as a submarine ridge. Bill Ryan's doctoral dissertation was centered on the theme that the eastern Mediterranean had been gradually lifted from the abyss to form the Mediterranean Ridge. This submarine ridge should ultimately become a mountain chain more imposing than the Alps. The evidence he needed to find was a sediment that could prove that the ridge had once been an abyssal plain. To a specialist, abyssal plain sediments are markedly different from normal sediments on submarine ridges. Coring of abyssal plain sediments in the northwestern Atlantic by American scientists, particularly by Maurice Ewing and Bruce Heezen of Lamont, has indicated that sands and muds deposited by turbidity currents find their way to the abyss of the ocean. These underwater currents can be described as ephemeral floods under the oceans. They can either come into existence as submarine continuations of unusually large river floods that empty into an ocean, or they can be generated by the sliding of sands and muds on a steep continental slope. The currents have a density greater than seawater because of sand and mud suspended in them, and they flow under gravity to the deepest depressions of the ocean floor, where the suspended sands and muds are dumped out as *graded* beds—graded according to the grain size, with the coarsest particles settling most quickly to the bottom. Eventually these graded beds fill up depressions, smoothing an irregular topography. This explains why the deepest part of an ocean basin is commonly a smooth plain called an abyssal plain.

Just as chicken-wire anhydrite is the trademark of sabkhas, graded beds are the sedimentary structure characteristic of turbidity current deposits. If the Mediterranean Ridge had been an abyssal plain before it was thrust up as a submarine mountain, we should find graded beds in the sediments under the ridge. Since the probable source for such turbidity currents lay somewhere farther south near the Nile Delta, the graded beds we wanted to find here should be composed of displaced desert sands and of black muds carried down by the Nile River. During his previous cruises Ryan had taken a number of piston cores on the Mediterranean Ridge, containing only normal oceanic oozes or skeletons of nanno-fossils and microfossils mixed with clays. None contained graded beds of black muddy sands from the Nile. This was not surprising, considering that the ridge is now more than a thousand meters above the abyssal plain and can no longer be reached by turbidity currents from the Nile. We could drill deeper with the *Challenger*, however, and look further back in time. If the ridge had once been an abyssal plain, we should find black graded beds. The age of the youngest graded bed on the ridge should mark the date when the sea bottom was about to rise from the abyss.

Site 130 was the pet idea of the chairman of the JOIDES Mediterranean Panel, Brackett Hersey, and Ryan enthusiastically supported the proposal. Their reasoning was logical enough, but there had been several elements of very theoretical deduction in their prognosis. During our planning meetings I had been most outspoken in my skepticism, although I reluctantly agreed to give the idea a try by drilling a quick hole in the Levantine Basin.

We arrived on location after a short cruise from the Strabo Mountain site. Drilling proceeded without incident, and the first core, at only twenty-three meters below the sea bottom, gave us our answer. Underlying the colorful pelagic oozes (the normal veneer of the ridge) we found black sands and muds, which could only have originated from the Nile. These were the turbidity current deposits that Hersey had looked for. We could now conclude that this part of the ridge had indeed been an abyssal plain before it was raised into a submarine mountain a million years ago.

The rise of the ridge is another manifestation of the collision between Africa and Europe. Whereas the sea floor basement plunged into the Benioff Zone under the Cretan Arc (as we discovered at Site 127), some of the sediments, because of their buoyancy, did not go down; they were sliced up and piled one on top of another (as we found at Site 129) to form topographic elevations, which might eventually reach the height of the Himalayas. Mountain building is a slow process, taking millions of years to accomplish, and different parts of the sea floor were deformed at different times.

Recuperating from my toothache, I slept through the night when the discovery was made. All day long on the seventeenth, however, I saw them bring in confirming evidence. Barrel after barrel of black Nile mud was hauled on deck. Late in the afternoon, Ryan woke up, and we decided to pull out of the hole to save some ship-time for our two remaining drill sites in the western Mediterranean; one was dear to Cita, and the other a favorite of Pautot's. We could not disappoint our friends. The drill string was raised above the mud line in the evening. Ryan wanted to take a few more cores near the sea bottom so as to provide a comparison with the piston cores taken during his previous cruises. Since he had been so correct in his proposal here, I was not in a mood to grudge him a few hours of ship-time. But he had no luck; the core catcher failed to catch any soft ooze, and we had three barrels of water to show for six hours of "digging" after midnight.

We drilled one more hole in the eastern Mediterranean, where we had thought the evaporites lay close to the surface. We had promised the paleontologists to make another attempt here to sample the sediments beneath the evaporites. Our prognosis on Site 131 turned out to be completely wrong, however. We got stuck in the Nile mud and never managed to get anywhere near the evaporites. Eventually oil drilling in off-shore regions of the Nile Delta did penetrate Messinian anhydrites, but they had to drill through more than 2,000 meters of sediments first—something we could never have done.

On September 19 we started our journey westward. We regretted that we had not gotten a good suite of evaporite cores in the eastern Mediterranean as we had in the Balearic Basin, but we did fulfill our primary objectives. Our cores told us that the Mediterranean had been a deep sea basin before it was desiccated in late Miocene, that the region had been under compression, especially during the last few million years, that the Mediterranean Ridge is rising, and that yesterday's abyss will form the mountains of tomorrow. The picture was now clear; we could no longer tolerate idle speculation that the Mediterranean was a shallow salt pan that foundered.

17

LADY'S CHOICE

—o—

THE MEDITERRANEAN is no longer a "Roman lake," but Italians would still like to think of the Tyrrhenian as their sea. This deep basin is dotted with underwater volcanoes, which we call seamounts (fig. 1), some of which rise above water to form islands; Stromboli is one such typical volcanic island. Other islands, like Elba or beautiful Capri, are not volcanic; they had been considered relics of a sunken continent.

Italian scientists had gathered some excellent marine geophysical and geological data on the Tyrrhenian in recent years, and during our meetings in Zurich Professors Morelli, Selli, and Cita-Sironi had shared some of their findings with us. Naturally they would have liked to have seen top priority given to some Tyrrhenian sites. But the French were lobbying for the Balearic at the same time, and the Germans for the Ionic. Our Italian colleagues were disappointed when the final proposal drafted by the Mediterranean Advisory Panel included only one top-priority Tyrrhenian site. The planning was perhaps adversely affected by the vagaries of circumstance. Professor Morelli had wanted to get together with Ryan and me in Trieste after our second meeting in Zurich to make concrete plans for drill sites in the Tyrrhenian. We were to consult his unpublished data during our visit. Unfortunately, the session had to be postponed when Ryan was unexpectedly called home. Then a second appointment was cancelled on the eve of my departure for Trieste in late April, when I received a telegram from the professor informing me that his university had been occupied by revolutionary students. At the last minute Ryan had to get some help from a colleague from Woods Hole, who made available to us a copy of the joint American-Italian survey of the Tyrrhenian. This information enabled the panel to pick our single site in the "Italian sea."

Site 132 was indeed an important one. We had planned to core continuously, hoping to obtain a second complete section of the Pleistocene and Pliocene sediments. Now as we sailed west, we had an added incentive to drill in the Tyrrhenian; this would help us evaluate further the two competing hypotheses. Had the Tyrrhenian also been a deep desiccated basin that had been overwhelmed by a sudden flooding at the end of the Messinian? Or had it been a shallow shelf sea that had foundered gradually after the salinity crisis? These questions were of great interest to us, and even dearer to the heart of Maria Cita. Ryan and I made a habit of teasing her; we would pretend that we might have to cancel the Tyrrhenian site for some more attractive alternative. Cita did not relish our humor, even though she knew that we had no intention of cancelling this "lady's choice."

Shortly after we departed our last site in the eastern Mediterranean, we ran into stormy weather. For three days the ship pitched and rolled. Stradner was down with seasickness, but some of us hardier souls steadied ourselves to work day and night on shipboard reports. Happily the sky cleared just when we drew close to the Strait of Messina. The captain had been a little apprehensive of trying to maneuver his vessel, with her tall derrick at midship, under the high tension wire that stretches across the strait. So we called in a pilot, who guided us safely through the channel, hugging the coast of Calabria. It seemed that we were close enough to reach our hands out to greet the pedestrians on shore. We

all recalled the story of Ulysses and his unhappy encounter
here with Scylla and Charibdis, but the olive tree that had
saved his life has long since been removed to make way for
the highway traffic.

In the evening of the twenty-third we dropped the positioning beacon for Site 132 southeast of Corsica. We did not
expect to have any problem obtaining a continuous record
of soft oozes here, for we almost always had one hundred
percent recovery when the core sequence consisted of ocean
oozes. We soon found out that our bad luck was still with
us, however. One of our toolpushers formulated that original
version of "Charlie's Law" here when I asked him why we
were not getting better recovery: "There are a hundred different reasons why you wouldn't get full recovery, and we
are learning them all."

Our trouble here turned out to be a plugged check valve,
a little orifice on the side of the core barrel. This tiny but
indispensable opening had been accidentally covered by a
plastic liner inside the barrel. We experienced many hours
of frustration before this small but not minor defect was
corrected.

Then when retrieving the twenty-fifth core at this site,
the crew had more difficulty and a minor accident, which
was amusingly reported by one of the amateur journalists
on board:

WHERE WERE YOU WHEN THE MUD HIT THE RAIL?

UPI dateline, 25 September 1970, Mediterranean Sea,
D/V *Glomar Challenger*: Coring at this important drill
site was nearing its end when a thin layer of hard mud
was encountered. Orders were given to core without a
plastic liner. After the barrel was hauled on deck, it was
noticed that the mud core adhered irrevocably to the inner
wall of the barrel and refused to budge. Core catcher was
dismounted and taken into the lab. Half-liners were taken

outside, and arrangements were made to pump the sediment out of the barrel. Water pressure built up slowly, 1,000, 2,000, 3,000 pounds . . . 4,000, 4,500, 4,750, 5,000! Suddenly, in a muffled explosion, a shapeless dark object flew in a low trajectory across the deck over the starboard rail. A soft haze of muddy water obscured the recoiling core barrel. Momentarily the front line workers scattered, but they quickly recovered and gathered the dispersed ejecta, carefully arranging the lumps in random order. Meanwhile the Chi-Sci rushed out, cursing mildly, and lamenting over the irretrievable loss overboard. The scatterbrained were chewed out for having failed to take the precaution of holding a bucket over the end of the mud cannon. Fortunately more lumps from the main deck and from the structural steel supporting the catwalk were recovered by enthusiastical personnel under the supervision of the Chi-Sci. Scientists and roughnecks joined forces and succeeded in a partial reconstruction of the approximately two meters of core by intuition, idiosyncrasy, and by gradation. Incident closed! Only a lone bystander, amateur painter by trade, bemoaned the fact that the newly refinished park bench (blue with black seat) was spattered by irrecoverable samples! Remarks: Next time when sticky sediment is being extruded and when the pressure approaches 5,000 psi, please notify the Chi-Sci immediately for precautionary bucketholding duties.

Despite this and other incidents, we got all the oozes we needed. Our most valuable find at this site, however, was a short core that told a fascinating story of the deluge.

According to this sample, the Tyrrhenian did not founder at the end of the Messinian; it was drowned. We found sabkha sediments here as we did in the Balearic, except that here the chicken-wire anhydrite had been changed into gypsum by ground water that had worked its way into this part of the Messinian desert. We found the dark marl laid down during the transient stage when the Mediterranean was being filled up. We found fossils of the last Messinian dwarf foraminifera buried in those black muds. At last we could put our finger on the sharply defined surface that recorded the instant when the dam at Gibraltar had broken, when the Atlantic water had poured into the Mediterranean, and when the Pliocene Epoch had begun! During the first thousand years of the new epoch the marine organisms were exclusively floaters or swimmers. Later, slow-moving bottom dwellers also managed to crawl across the strait. So Dick Benson could now his find benthonic ostracods and Orville Bandy his benthonic foraminifera in the sediments a few centimeters above the sharply defined surface that signaled the beginning of the Pliocene.

Studies of microfaunas in the cores provided a picture of the changing ecology of the Mediterranean. Apparently the Strait of Gibraltar was being gradually shoaled during the Pliocene so that deep and cold Atlantic waters were eventually barred from the Mediterranean. The bottom waters of the Mediterranean were gradually warming up, and the cold-water creatures among the bottom dwellers died out; the last of the Mediterranean's deep-sea dwelling ostracods perished some 2 million years ago. Ventilation of the bottom waters was only possible when surface waters chilled by winter air over the Balearic and Adriatic seas dropped to the bottom. In fact the situation got so bad during the Pleistocene that the whole eastern Mediterranean sea floor became repeatedly stagnant; only stinky black dirt, called "sapropel" by geologists, was deposited on a sterile Mediterranean seabed. The sill of the strait was apparently getting shallower every year. It was getting harder and harder to repair the damage of excess evaporation in the Mediterranean. Nevertheless a rather efficient disposal of the heavy brines through a reflux of dense evaporated water from the Mediterranean back to

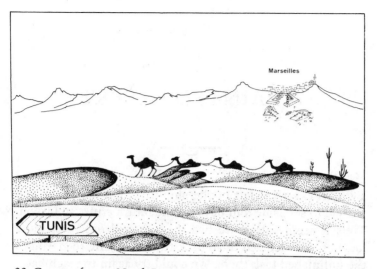

33. Cartoon from a North Borneo newspaper that reported the findings of the Leg 13 cruise.

the Atlantic prevented the Mediterranean from turning into a heavy brine pool. But the days of the beautiful blue sea are numbered. History will likely repeat itself. One can imagine in the not too distant future, say some two or three million years from now, that Gibraltar will be again changed into an isthmus. The Gibraltar falls will be rejuvenated. Cruise ships will be replaced by camel caravans across the new Mediterranean desert. Huge hydroelectric plants will be built beside the saltwater fall to ameliorate the ever growing energy crisis. Oil derricks will spring up to exploit the giant oil fields under the newly desiccated playa bottom, where the petroleum reserves may surpass even the richest of the Middle East. Only the Riviera and Costa Blanca will suffer a decline; they will be deserted except for a few roughnecks who might seek R and R in those desert outposts.

18

A PAINTED DESERT

———o———

TWO WEEKS were left of the ship-time allotted to us, and we had to set aside four days for the return cruise to Lisbon. Thus it seemed that we had time to drill only one more location. We decided to go back to the Balearic to attend to the unfinished task there. We could try again to reach basement near Site 124 on the western margin of the Balearic abyssal plain, but Pautot told us that there was another "basement high" on the eastern margin. We might as well drill our next hole in the eastern Balearic.

We chose for Site 133 a place where we could drill into the M-reflector about 100 meters below the sea floor and touch basement some 200 meters down. We expected to core soft oozes and drill through a thin evaporite formation before sampling the basement—our primary objective.

The second core brought us a surprise, however. We had drilled into the M-reflector expecting anhydrite, or gypsum, or dolomite, or some other evaporite mineral in our core. But when the plastic liner was split open, we found instead well-rounded gravels between massive red and green silts. What were *they* doing here? We drilled deeper and took another core. More red beds. We drilled deeper yet and took another core. More gravels and more green and red silts! None of these sediments contained fossils of any kind; no foraminifera, no nannoplankton, no diatoms, no ostracods, no clams or snails. Finally it began to dawn on us that we were drilling into an ancient desert creek bed!

During our three-day voyage eastward to the Ionian Basin, shortly after we had first entertained the idea of a deep desiccated Mediterranean basin, Ryan, playing devil's advocate, had tried to punch holes in our new theory:

"No, the evidence does not suggest that the Mediterranean basin dried up. Rivers draining into such a basin would bring down a lot of sands and gravels. Where are they? We have nothing but anhydrite in our cores."

"We haven't recovered any sands and gravels because our drill sites are too far away from the coast. Flash floods coming down an arroyo might build an alluvial fan at the mouth of a canyon, but you can't really expect any of these ephemeral streams to bring sands and gravels to our salt pan."

"But sands and gravels do get into the heart of Death Valley. There are the dunes near Stovepipe's Well, and there are the gravels under the Devil's Golf Course."

"Yes, but Death Valley is so narrow compared to the Balearic abyssal plain. We might get some windblown sands here and there, but we can't count on any desert stream to flow hundreds of kilometers across a flat basin floor."

My argument made sense. Still, Ryan was not entirely satisfied; he wanted to see signs of a painted desert. Now, by chance, we had stumbled onto the right location, in Hole 133. We were 160 kilometers west of Sardinia, at the foot of the continental slope. The continental slope here had become a mountain front when the seawater had been removed by evaporation from the Balearic Basin, and at the foot of this escarpment the Messinian desert creeks had built their alluvial fans (fig. 34). I remembered the manuscript by Hardie and Eugster, which described arroyo gravels and red beds rimming the Sicilian salt basins. Here we had the same kind of gravels and the same kind of red beds.

Ryan, for his part, began to recall the gravels dredged up some years ago by Bourcart from submarine canyons in the Mediterranean. Apparently the French had been busy exploring the underwater topography of the western Mediter-

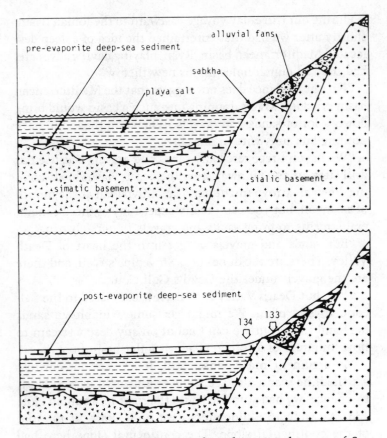

34. Schematic diagram showing the paleogeography west of Sardinia in late Miocene (upper) and early Pliocene (lower) epochs.

ranean during the decade after the Second World War when Bourcart and his associates found many of these submarine canyons. The Mediterranean canyons seemed to be different from those found on the continental margins of the Atlantic and Pacific, however. They appeared to be drowned river valleys, whereas the Atlantic and Pacific canyons appeared to have been cut by submarine turbidity currents. Furthermore, many of the canyons off the Côte d'Azur had not eroded recently, or during the Pleistocene, as the Atlantic and Pacific canyons had; they had been cut during the late Miocene. They were partly filled with late Miocene river gravels and then covered by ocean oozes of Pliocene age (fig. 35). The heads of many of these large submarine canyons could be linked to the mouth of modern rivers in southern France, Corsica, Sardinia, North Africa, and Spain. The bottoms of the canyons could be traced to about the level of the Balearic abyssal plain.

The origin of the canyons and the gravels had constituted a puzzle. Bourcart was convinced that the canyons had been cut above sea level by late Miocene streams. Not aware of any good evidence to suggest that the Mediterranean might have dried up, he proposed the less outrageous hypothesis that European and African continental margins had been bent down, drowning the Miocene coastal streams. We suspect that our French colleagues on Leg 13 were still much influenced by their late master's hypothesis of "peri-continental flexure" when they postulated a post-Miocene foundering of the Mediterranean. But Bourcart's hypothesis was never very satisfying. Coastal streams would not make canyons. And streams leave their gravels in the mountains. The late Miocene gravels in the Mediterranean submarine canyons could only have been deposited by mountain streams, not by lazy meandering rivers on flat coastal plains.

As we sat in the core lab admiring the red and green desert sediments, we saw a new explanation to Bourcart's findings. The Mediterranean had been dry during the late Miocene, and we could envision a painted desert at the bottom of the present continental slope, stretching across the wide expanse of what is now the Balearic abyssal plain. The desert floor then lay more than 2,000 meters below the level of the sea on the other side of the Gibraltar. Rivers in circum-Mediterranean lands were no longer emptying into an inland sea at sea level. Instead, they had to run a steep course down the newly exposed continental shelf and slope. Old coastal plains were now perched high and dry on the margin of high plateaus rimming the desiccated Balearic. Rejuvenated streams made deep indentations on the edges of these plateaus and sculptured grand canyons on their way down to the dried up abyssal plain. Gravels were dumped in the canyons and variegated silts were piled up on alluvial fans at the foot of the

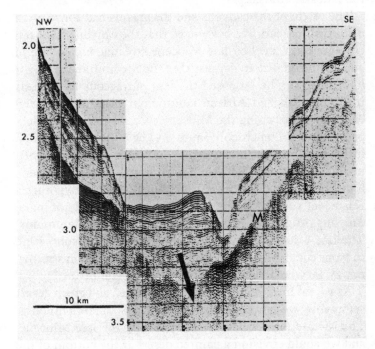

35. The reflection profile between Sardinia (left) and Tunisia (right) shows an excavation in horizon M by stream erosion during the time when the Mediterranean dried up. The modern valley was cut by submarine turbidity currents. This seismic record was made in 1968 by M/S *Amazon* and has a vertical exaggeration of about 30 to 1.

escarpment. With this hypothesis, we not only explained the occurrence of red silts at Site 133 but provided at the same time a neat answer for Bourcart's canyons and gravels; we also resolved the long standing mystery of the down cutting of the Rhone in southern France.

19

THE BULL'S-EYE

—O—

WE WERE elated with our new discovery, but Anderson was gravely concerned with a new difficulty on the rig floor. The red and green silts were poorly consolidated, and our drill hole had no casing. The loose debris could thus slump easily down hole, and we could not pump hard enough to wash out the silts. The inevitable finally struck when we cut the seventh core. We had to shut off the pump temporarily, the wall collapsed, and we wound up with another stuck drill string. It was the same old story. We pumped in some mud and applied tension on the string. Ryan slept, while Anderson asked me to pick a new location.

After a couple of hours, the string was pulled free. I wanted to drill ahead at the same location. After all, we still had not reached our basement objective. Anderson said no. We had an argument and were both pretty upset, so we decided to wake Ryan up for consultation. There was no time to lose. The roughnecks were waiting, and a decision had to be made immediately. Should we risk getting stuck again in the silts, or should we pull out of the hole while we still could? There was ship-time enough to drill another hole for our basement objective, but we might never reach basement if we stayed here and got stuck again. Anderson's last argument made sense, and Ryan, half awake, gave in. I was furious. Again we would have to abandon our hole before we could reach our primary objective. I could not know then that we were to hit the bull's-eye in our next hole.

"Bull's-eye" was an expression used by Bob Schmalz to describe the distribution of playa evaporites. He recognized two different patterns for two different modes of evaporite deposition. If a suite of evaporites had precipitated out of a deep brine pool maintaining a restricted communication with the open ocean, the evaporite distribution should appear on map in a teardrop pattern; the most soluble salts, and the last salts to settle out of a bittern, should appear at the end farthest away from the pathway to the open ocean. If, on the other hand, evaporites were residues paving the floor of what had been a completely isolated basin, the first salt to precipitate on the periphery would be carbonate—a limestone or dolomite. As the water level dropped and the brines became more concentrated, a ring of sulfates would settle out. Finally in the center or deepest depressions of a salt pan, we should find the bull's-eye, where halite and other more soluble salts would have been precipitated (fig. 36).

Our drilling results so far permitted us to continue contemplating a bull's-eye in the Mediterranean. We had found sulfate deposition on basin margins slightly above the abyssal plains (fig. 37). And our seismic records indicated that there must be rock salt under the central abyssal plains, for we were certain by now that the array of pillarlike structures could not be anything but salt domes. All we needed to prove our interpretation was to obtain a salt sample. The trouble was that we could not drill on a salt dome because of the probability of encountering an oil reservoir and the ensuing danger of pollution. Where we *could* drill we could not expect to get through to the salt. To discourage us, Anderson swore that we would never take a salt core with our primitive equipment, even if we drilled straight into a salt bed. He was sure the salt would be long gone before we could raise the core barrel, being dissolved by the seawater pumped down

the hole as circulating fluid. Ryan and I had felt very frustrated throughout the last few weeks. We were almost sure
we had a bull's-eye there—one more proof that the Mediterranean had dried up. We wanted very much to bring up
a piece of rock salt, but we could not permit ourselves enough
optimism to plan for such an operation.

In the late morning of the twenty-ninth Anderson left us
alone in our cabin to choose our next, and last, drilling location. Our drilling results at Site 133 had indicated that the
"basement high" was not a volcano as we had thought, but
a "basement ridge." The seaward side of this ridge was buried
under a thin wedge of abyssal plain sediments (fig. 34). Time
was then getting short, and I was still obsessed with getting
down to basement. So I proposed that our next drill site be
located near the top of the ridge where the sedimentary cover

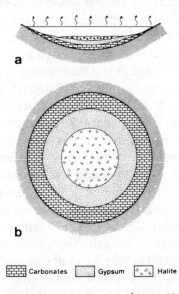

a

b

| Carbonates | Gypsum | Halite |

36. Bull's-eye pattern of evaporite
deposition. Courtesy of DSDP.

was thin. Ryan made a counterproposal that irritated me to no end; he wanted to drill farther seaward at the foot of the buried ridge where the basement was covered by at least several hundred meters of sediments.

"We could not possibly get to the basement under the thick abyssal plain deposits there." I was exasperated.

"I know; but we have a few days of ship-time left. We don't need to panic. We can reach basement on top of the ridge anytime we want to. But we still have time to drill a hole in the basin before making an offset on the ridge. I want very much to find another record of the Pliocene deluge here in the Balearic. We missed it when we drilled Hole 124."

"We don't have time for that. If we want to make sure that we hit Miocene-Pliocene contact, we have to do continuous coring. We simply do not have that kind of time!"

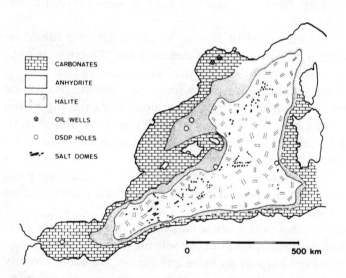

CARBONATES

ANHYDRITE

HALITE

⊗ OIL WELLS

○ DSDP HOLES

〰 SALT DOMES

0   500 km

37. Map of inferred distribution of the Mediterranean evaporites in the Balearic Basin, in accordance with the postulate of complete desiccation.

"It's worth a try, Ken." Ryan was now trying all kinds of persuasion to win his case: "The drilling crew is nervous and jittery now, having run into so much trouble lately. If we drill on the buried ridge where the sediment cover is thin, we might reach basement. But the bottom-hole assembly could get stuck in the hard rock and be twisted off before the hole could be stabilized. Then everybody would be upset. We have to do something safe now to soothe their nerves and boost their morale. If we spud our next hole in the basin, we won't be getting into any trouble. They could easily slip through several hundred meters of section and they would have a sense of achievement. Meanwhile, the drill string can be stabilized and we won't get stuck. We could then try the ridge top offset. We'll have plenty of time to do that before we have to leave for Lisbon."

"Okay, you win. We'll try the basin hole," I gave in unwillingly. "But we absolutely have no time for continuous coring."

"We don't need to anyway until we get near the M-reflector a couple of hundred meters down. We can wait until our seismic record tells us that we are getting near."

"But you know now how lousy your seismic is. We are never sure where we are. If we don't miss the top of the reflector by a hundred meters, we'll miss it by fifty at least. And even that will mean six barrels of continuous coring for nothing; we'll be wasting our time."

"No, we'll make the contact in five hours. Anyway, there is no harm in giving it a try. You will have plenty of time for your basement hole."

So another compromise was made. We went to the electronics lab and spread out our seismic charts. Pautot's *Charcot* survey gave us an excellent record. We could identify the M-reflector. We could even recognize a seismic reflecting horizon, which signified a layer of rock salt. We noticed that this layer lay only a few hundred meters below the sea floor at Ryan's proposed location, but neither of us ventured to mention it as a coring objective. We did not believe that we would be lucky enough to get down there before it was time for us to quit again. Or perhaps we did not want to alarm Anderson prematurely.

Anyway, a new location was picked, and its coordinates were communicated to the captain. I wrote a new prognosis announcing the site to our shipmates. They were upset that we had to move again, but they were too tired to protest. It was only a short ride, and we arrived at Site 134 at 1705 in the afternoon.

We had a very frustrating beginning at our new location. First we had difficulties again with Anderson. The disagreement was over a technical triviality, but his insistence on doing everything the extra-safe way cost us a couple hours of valuable ship-time. We had been very lucky so far and had lost no equipment, but we began to suspect that Anderson was more concerned about keeping a clean slate on that score than about scientific objectives. He, the drilling superintendent, and the toolpusher seemed to be fooling around here; and the crew did not get set to spud the well until after dark. Once we finally got started, we drilled through the soft stuff without any difficulty. But after we had drilled only 170 meters, the needle on the pressure gauge flicked, a sign that we had hit hard rock. This was about a hundred meters short of the M-layer, as Ryan had calculated it. He immediately decided to take the first core, thinking we might have misidentified the M-reflector.

Thus we started coring operations sooner than planned. All evening long people asked me when we expected the first core to be up. I kept saying: "Soon—in about two hours." Late in the evening I quit guessing. The core came in after midnight, three or four hours later than we had expected, and all we had was a barrel full of water. By then Ryan had gone to sleep for a couple of hours; I asked the toolpusher to cut another core immediately below. Ryan relieved me at about two o'clock in the morning. When I went down to breakfast at six, Ryan told me that he had had a terribly frustrating night. Nothing seemed to work, and we had recovered practically nothing. He had such a backache now that he had to get some rest again. Backaches always seem to get worse when things go wrong.

So I took the logbook and went well-sitting. It was a beautiful morning. The sea was calm and the wind was still. Jim, the toolpusher, had promised for some time that he would take me up the elevator to the top of the drill tower. That morning I finally put a roll of film in my camera and took him up on his offer.

It was an open-air lift. One almost had the feeling that one was going up the Eiffel Tower. And the view from the top was glorious. Stacked neatly below at the bow were rows and rows of drill pipes. Looking astern into the sun, I saw the bridge and our living quarters. Some early risers were already on the deck enjoying the morning sun. On my way down I was given a demonstration of the esteem and affection that the roughnecks held for their chief scientists. I was to be lifted from the elevator of the derrick tower by a crane and lowered down to the rig floor. But the crew left me suspended in midair during the transfer. Furthermore, they threatened to train water hoses on me to give me an early morning shower. Jim, the old reliable, also joined in the fun. When I was finally let down, I had to scowl at him and reply, "Et tu, Brute!"

Soon we had our fifth core—and our first success. We actually found some mud in the barrel. From this, Cita told us that we were still in the Pliocene, quite a distance above the top of the evaporite series. All morning we had proceeded cautiously. We did not have enough time to core continuously. On the other hand, we did not dare move too fast lest we miss our contact. Things went along fine for awhile; we got some cores, but I kept revising the depth at which we expected to encounter the M-reflector. The drillers poked fun at me again, suggesting that I start a lottery.

Shortly after lunch, Ryan woke up and inquired about our progress. He was at least consoled that we were bringing in cores, even though his optimistic appraisal of meeting the M-reflector in five hours proved completely unrealistic. Almost twenty-four hours had gone by—precious hours had been reserved to investigate the basement below the evaporites, but we had not even reached the evaporites. Worse still, we began to encounter sand layers. Perhaps the reflectors on the seismic record were in fact not layers within the evaporite formation at all, but cemented sandstone in a Pliocene abyssal plain sequence. Ryan and I began to worry that we had completely misread our seismic record.

Since I was short on sleep, I should have left the well-sitting duties to Ryan, but I was too keyed up to go to bed. I had read all the old paperbacks on the shelves in the science lounge, so I went to the science office to read old cruise reports as an exercise in relaxation. I took out the two thick volumes prepared by Jerry Winterer and his crew when they went to the Marianas in the Pacific.

After a couple of hours I was ready for bed. But since it was just about time for another core to come up, I thought I might as well go to the rig floor to take a last look. Coming out of the science office, I bumped right into Ryan in the gangway. He had a shining icicle in his hand. "Taste it," he exulted. "It's salty. We hit rock salt!" It was a complete surprise for all of us.

Ryan led me to the core lab. The room was full of people— scientists, marine technicians, roughnecks, sailors, even the cook. All had come to admire the salt core. It was indeed partially dissolved by circulating seawater as Anderson had predicted, but we had hauled it up!

Since all of the scientists were in one room, perhaps for

the first time during the cruise, we decided to have group pictures taken to commemorate this historic occasion. Someone woke up Orrin Russie, the photographer, and we all got set posing for publicity. The captain was called in. They also found Anderson. Just this moment Jim, the tool-pusher, came in and asked me quietly:

"How much did we get?"

"Salt!" I said, still beside myself.

"No, I mean how many meters of cores did we get?" The toolpusher had to keep record of meters cored and recovered in his book.

"Oh, I don't know. Put down 0.5 meter for your record, but it is worth more than a thousand meters of mud!"

This was the first time that rock salt had ever been recovered from the ocean floor, and we had hit the bull's-eye in the Balearic 3,000 meters beneath the sea!

20

HOMEBOUND IN A MISTRAL

—⊖—

AFTER the salt, the rest was anticlimactic. We thought we had missed the Miocene-Pliocene contact we were looking for because we had not cored continuously, but we were happily surprised in the evening of the thirtieth when Cita reported that she had just found another "silent witness of the deluge" in a core that had been brought up earlier and was only then being opened. So the Balearic, too, had been suddenly flooded 5 million years ago when the dam at the Gibraltar had burst.

We also found a sample of an ocean ooze between two salt cores that was made up largely of tiny skeletons of deep-water foraminifera. Rock salt had crystallized out when the catchment basin dried up, but we now had evidence that there had been a deep sea in the Mediterranean between episodes of desiccation. Cita identified the fauna as latest Miocene, indicating that the Balearic had been alternately an inland sea and a playa lake during the Messinian age, at about the same time the Tyrrhenian and Ionian basins were also periodically dry.

We would have liked to have dug to the bottom of the salt, but we had to quit. The drilling rate had again slowed to a snail's pace, but more alarming was the fact that the ooze sample, sealed between the layers of rock salt, smelled strongly of gasoline. If we continued to bore, we might easily hit oil. Reluctantly we raised the drill string to the deck and cemented the hole. We then drilled an offset hole on the ridge top.

There was no problem sampling the basement under its thin sedimentary cover here. We actually managed to obtain cores of basement rocks in a series of five offset holes. There had to be much maneuvering, and Captain Clarke was most helpful; he was always on the bridge in command when an offset had to be made. During the last sixty hours of our operations, neither Ryan nor I slept a wink, and the captain hardly much more. Finally they insisted that we bring in the drill string and leave for Lisbon. Ryan and I, intoxicated by our success, wanted to go for one more offset. We were told that it was not possible. There was a long voyage home, and the ship had to slow down and idle for half a day en route while the crew cleaned and tied up their equipment.

Reluctantly, then, we followed orders. At midday, October 2, we watched them begin operations to pull out of our last hole. Cita, Maync, and Wezel brought their cameras, for they had not been able to find much time for photography, and now was their last chance to make their pictorial record of the "ballet" on the rig floor (see fig. 24). We were somewhat upset with Anderson that afternoon when we discovered that we had quit a few hours early to give him a chance to try out a piece of new coring equipment. Constant bickering went on between the co-chief scientists, who wished to invest every minute of ship-time in pursuing scientific objectives, and the operations manager, who worried about his equipment and technical aims.

Actually we did not leave at all too soon. The captain ordered the start of our final westward journey at 2000 on October 2; and just as the crew had hauled up and secured the bottom-hole assembly, the mistral began to blow.

I could not get to sleep at the start of the storm, so I went looking for my colleagues. I found Cita, Maync, and Ryan in the paleo lab gathered around a pile of cores, preparing samples for on-shore specialists. They were digging out samples, labeling vials, and sealing the samples in the vials with masking tape. The whole operation reminded me of a big Chinese family making won tons in their kitchen. It was a tedious job, but just what the doctor ordered for insomnia, so I joined the "won ton" production line.

As we sat reminiscing, it seemed that we had had nothing but trouble, starting with the pilot boat running out of gas in Lisbon. There was the Dumitrica story and the threat of an order to return to Gibraltar. Then we got stuck—in gravels, in sands, in volcanic ashes, in gypsum, in red silts. When we were not stuck, we were spinning our wheels in waxy shales or grinding away at the "pillars of Atlantis." We were always choosing the wrong drill bit, but if we did choose the right one, it would be worn out just before we reached basement. The cruise seemed to be one long nightmare. Yet we had also had more than our share of joy and excitement—when we cored ophiolite just before we were to call a quit, when we first washed out the unusual gravel and found the shining gypsum, when we first brought in a "silent witness of the deluge," when we met tiny mollusklike creatures who had lived only in brackish lakes, when we found black Nile muds, and finally when Ryan waved that glistening salt core in his hand. We could persuade ourselves that the money from the National Science Foundation was well spent.

We finished the vial samples and had our usual midnight snack. Then I went to the science lounge and tried to draft the operations narrative for our last site. Because of the storm I soon began to have difficulty sitting up straight, however, and a filing cabinet in the corner of the lounge started making a racket as its drawers opened and shut with the roll of the ship. I had the beginnings of a headache when I went back to my cabin.

I must have slept more than twelve hours, even though all night the file cabinet in our room beat out a slow rhythm. More than half asleep, I still was conscious enough to grasp the edge of the bed. Ryan apparently had to fight even harder to stay in his upper bunk. Still, he slept, if not too soundly. We were very, very tired. Now that the drilling operations were completed, we could relax and sleep through the storm.

When I went to the mess hall in the early afternoon of the third, it was almost deserted. We did not have steaks for lunch, because the cooks were sick, so I took a roast beef sandwich and a flask of coffee to the bridge. The third mate was on watch. The storm was dying now, he told me. During the night the roll had exceeded 35 degrees; never had it been so bad on the *Challenger*, not even when they had encountered a typhoon in the Pacific.

The storm died down altogether the next day, and the sun came out when the crew started cleaning and tidying things up. The scientific staff also had to work, for we had shipboard reports to complete before we arrived at Lisbon. We held a lengthy staff conference to decide who would do what part of the final cruise report, which would be known, ironically, as the Initial Reports of the Deep Sea Drilling Project. Then we broke up to begin our writing. In the evening I started to work on the summary of our shipboard opus. I did not finish until the early hours of the morning. But then I could not sleep, despite two sleeping pills and hours of reading the Bible. I got up several times to write. Ryan, who had gone to sleep early in the evening, finally offered me the last of his whiskey to help calm my nerves.

Finally, I slept. But at seven o'clock the next morning, the third mate came in to wake me up. The captain had started his homeward journey too early, and now he wanted to kill some time so that the *Challenger* would not dock before the contractual date. They wanted to use some ship-time to make seismic surveys. I was furious. Later I had a very unpleasant session with the captain. We needed sleep more than surveys, I said, and he would not have had to search for ways to waste ship-time if he had given us the last offset.

Soon I was to regret my outbursts, however. After all, we had worked harmoniously together for almost eight weeks. Tempers flew everywhere during the last day of our voyage. The pressure was heaviest on our secretary, Sue Strand, who had to finish typing the whole shipboard report. Sue came on board during "Operation Broadhaul" to replace Eleanor tum Suden, who had worked as a "yeoman" during the first four weeks when we had not gotten around to much writing. Eleanor has a nice way of pleasing people while leaving things undone, thus poor Sue was driven to near martyrdom trying to catch up. The bulk of the shipboard reports came in during the second half of the cruise, and the volume now filled three thick folders. On this last day all of us ran to Sue with our piles of manuscripts to be typed. Needless to say, she did not sleep again until we reached Lisbon; but our report was typed and ready for distribution when we docked.

All day long on the fifth, people ran around tying up loose ends here and there. That evening I stayed in my cabin drafting explanatory notes to accompany the exhibit specimens for the press conference to be held after we got back. But Cita looked in and asked me to share her last bottle of port with her and the paleo crew. So I went down to the paleo lab with her, wetted my lips, and indulged in a last round of reminiscing. To conclude the celebration and the cruise, Maync read a ditty that neatly summarized our trials on *Glomar Challenger*:

> R 0100 06 10 70
> FM Microscopic Lab
> To Chief Scientists, D/V *Glomar Challenger*, WNCU
>
> To Bill Ryan and Ken Hsü, our Beloved Chiefs
>
> One is against, the other pro,
> "Maybe," says one, the other "No."

Thoughts and ideas conflict and clash,
For human brains work like a flash.
Concepts different do arise
Of smaller or a larger size.
To settle such discrepancies,
Questions, and divergencies,
Cavemen are apt to use their fists—
But not our two chief scientists!
They explain, dispute, and discuss;
In short: They make a real fuss.
A subtle design to open the blind
So primitive and naive mind!

To punch a hole into the crust
Is for those two a simple must.
Worldwide tectonics, structures rifted,
Split and turned, continents drifted.
Abyssal plains and trenches deep,
Subsiding troughs with walls so steep,
The hidden crust, the ocean floor:
They want to drill and wish to core.
The basement which no one has seen
(Sometimes it's only Miocene!)
"By gosh, let's core! That's real stuff,
Below there lies a hidden trough
Slice forth we must, it's more than worth,
The bowels of old Mother Earth!"

Reflector "M" is quite a pet
To Bill and Ken, we sure do bet,
But what a deal means drilling through!
They know by now, Ryan and Hsü!
Basalt, tephra plus andesite,
Bring to our chiefs just sheer delight.
When they should get a few small chips,
A radiant smile transforms their lips!

The soaring hopes, they bring no luck:
This hell of a bit gets often stuck!
The waxy shales are much too hard
(Although they look like melted lard!)
Limestones and dolomites are well known
As rock impervious to drill down.

The evaporites are all very tight:
"Let's go and drill another site."
Of sediments they dream and muse,
Of sapropels and nanno ooze;
If sands look neither wrong nor right
They simply yell: "A turbidite."
But now enough of making fun!
We thank you today for all you've done,
For all your effort, day and night,
That you displayed on every site.
We want to thank you in this way,
And herewith wish you both today
A future happy as a dream—

—The Foram-, Rad-, and Nanno team.

EPILOGUE

—o—

It is an old maxim of mine that when you have excluded the impossible, whatever remains, however improbable, must be the truth.—Arthur Conan Doyle, *The Lost Coronet*

THE *Glomar Challenger* returned to Lisbon at 0800 on October 6, where we were greeted at the pier by Mel Peterson, the project's chief scientist, and his retinue. After two days in Lisbon, where we were treated like heroes returning from battle, we were herded to Paris. Scripps had arranged a press conference there through the French National Center of Ocean Exploitation. Bill Nierenberg, the project's principal investigator and the director of Scripps, and Dan Hunt, National Science Foundation representative, joined us there. We told the press the story of how we had discovered evidence of an ancient Mediterranean desert 3,000 meters below sea level, a story that made headlines all over the world. Then we settled down to the tedious work of writing our "initial" cruise report.

We envisioned the Mediterranean 20 million years ago as a broad seaway linking the Indian and the Atlantic oceans. With the collision of the African and the Asiatic continents and the advent of mountain building in the Middle East about 15 million years ago, the connection to the Indian Ocean was severed. Meanwhile, the communication to the Atlantic was maintained only by way of two narrow straits, the Betic in southern Spain and the Riphian in North Africa. We saw evidence in our cores of the gradual deterioration of the Mediterranean environment: the advancing stagnation of its waters, the inevitable extinction of its bottom dwellers, the struggle for existence by its swimming and floating population, and the evolution of a hardy race that could survive the salinity crisis. With the final closure of the two straits, the inland sea then became a series of great salt lakes, which became desiccated from time to time leading to the complete extermination of the fauna and flora at the bottom of this Miocene death valley, 3,000 meters below sea level. The isthmus of Gibraltar was the lone barrier between the desiccated Mediterranean and the invading waters from the Atlantic. When the dam broke, at the beginning of the Pliocene 5 million years ago, seawater roared through the breach in a gigantic waterfall. Cascading at the rate of about 40,000 cubic kilometers per year, the Gibraltar Falls were one hundred times bigger than Victoria Falls and a thousand times grander than Niagara. Even with such an impressive influx, it took more than one hundred years to fill the empty Mediterranean. What a spectacle it must have been!

During the Messinian salinity crisis, faunas came whenever seawater spilled over the isthmus and died out when dissolved ions were precipitated as salts. But a new faunal dynasty was established after the deluge. At the beginning of the Pliocene, the strait was narrow but deep, and cold Atlantic waters could find their way easily into the newly reborn Mediterranean Sea. But our cores indicated that history may be repeating itself. The Strait of Gibraltar is being gradually shoaled to become a shallow sill. We can envision the Strait of Gibraltar again being an isthmus between the Atlantic and a Mediterranean desert.

These were tall tales. But Ryan, Cita, and I preached the story as the rocks directed. Our somewhat outrageous idea was naturally greeted with disbelief by many. Questions were raised and criticisms made. Yet each question-and-answer session only turned up new pieces for our jigsaw mosaic.

When we first stepped down the gangplank after the *Challenger* returned, we had only identified gypsum, anhydrite, and halite in our suite of evaporite minerals. We were everywhere faced with the embarrassing question of where the carbonate was, the first mineral to come down from an evaporite brine. We thought we might use the idea of saline mineral zoning as an excuse; perhaps our drill holes all hit the regions where only sulfate minerals would precipitate. It was a weak argument. The Alboran hole was high on the periphery of the Mediterranean; we should have found evaporative dolomite there. So we took out the late Miocene cores from Site 121 again. Sure enough, we found dolomite here where we had failed to recognize it before. But it would be nice to see more. Our hopes were fulfilled in a few months, once Vlad Nesteroff had had a chance to study in leisure the core samples he had taken back. He found the very fine sediment from Hole 124 that we had not been able to identify on board to be made up of a dolomite mineral.

We were also asked why, if the basin had dried up, there were not more soluble potash and magnesium salt in our cores. We could counter with the argument that we had only hit a corner of the bull's-eye, not its center where the last bitterns would accumulate. The argument made sense, but it did not satisfy the skeptics. Two years after returning from the Mediterranean I finally met a specialist in salt chemistry, Robert Kühn of Kaliforschung at Hannover, who kindly consented to do a detailed chemical analysis of our salt samples. And indeed, he found bischofite, one of the more soluble magnesium salts that we should find in a desiccated playa. Further, he performed a series of trace element studies whose results firmly supported our conclusion that the Mediterranean salts had been deposited in shallow salt ponds, not in deep briny lakes.

During one of her talks in Lyon, in a symposium attended mostly by micropaleontologists, Cita was asked what geomorphic evidence supported the desiccation of the Mediterranean. If the Mediterranean had indeed been emptied of water, the coastal plains of the surrounding lands would have become high plateaus, and islands would have been lofty peaks. The first response to a lowering of the water level would be a rejuvenation of streams and a marked increase in their erosive power. Cita was not a specialist in such matters, but a French colleague, G. Clauson, rose to her defense. He told the audience, and explained later also to us, the story of the down-cutting of the Rhone, and Denziot's prophetic interpretations. There were other rivers draining into the Mediterranean, too, however. They too must have cut gorges? Where were they?

Ryan was soon to come up with an answer to that question. Shortly after we returned to port, he received a letter from a Russian geologist, I. S. Chumakov, who had learned of our findings through an article in the *New York Times*. Chumakov was one of the specialists sent by the USSR to Aswan to help build the famous high dam. In an effort to find hard rock for the dam's foundation, fifteen boreholes were drilled. To the Russian's amazement, they discovered a deep, narrow gorge under the Nile Valley, cut 200 meters below sea level into hard granite. The valley had been drowned some 5 million years ago and was filled with Pliocene marine muds, which were covered by the Nile alluvium (fig. 38).

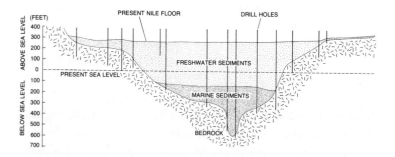

38. Deep gorge under the upper Nile Valley near Aswan, discovered by a team of Soviet geologists while drilling test boreholes preparatory to the construction of Aswan Dam.

Aswan is about 1,200 kilometers upstream from the Mediterranean coast. In the Nile Delta, boreholes more than 300 meters deep were not able to reach the bottom of the old Nile canyon. Chumakov estimated that the depth of the incision there might reach 1,500 meters, and he visualized a deeply buried estuary under the sands and silts of the modern Nile Delta. Chumakov was right; a narrow 2,500-meter-deep canyon under Cairo was recently discovered during geophysical explorations for petroleum in Egypt.

Chumakov was not the only one who found buried gorges. Petroleum geologists exploring in Libya described their surprises. First, their seismograms registered anomalies: there were linear features underground transmitting seismic waves at abnormally high velocities. Drilling into the anomalies revealed that they were buried channels incised 400 meters below sea level. The geological record tells the same story: vigorous down-cutting by streams in the late Miocene and sudden flooding by marine waters at the beginning of the Pliocene. Ted Barr and his coworkers in the Oasis Oil Company, based in Tripoli, Libya, concluded in a report that the Mediterranean Sea must have been a thousand meters or more below its present level when the channels were cut. They could not get their manuscript published in a scientific journal since no one would accept such an outrageous interpretation.

Still other buried gorges and channels have been found in Algeria, Israel, Syria, and other Mediterranean countries. So the evidence on land is plentiful to support our story. But in the course of a lecture at Yale University devoted to a discussion of the results of the Leg 13 expedition, a student in the audience asked for confirming evidence from the sea. The rivers draining into the Mediterranean should not only have incised deeply on land, they should have continued on down across the exposed continental shelf and the continental slope to the flat bottom of the abyssal plain, which was then turning into a playa. Where are such channels? Ryan had come to New Haven to attend my lecture, and he could now back me up. He told the audience of the extensive oceanographic surveys by the French in the Balearic Basin and of the enormous submarine canyons they discovered. These canyons, as Ryan explained, are typically river cut and are filled with river gravels. Most of them can be related to a river on land and can be traced to a depth of about 2,500 meters at the edge of the abyssal plain. They too were drowned by the early Pliocene deluge. Similar canyons have been found in all parts of the Mediterranean. Their genesis had never been satisfactorily explained until we found evidence that the Mediterranean had dried up 6 million years ago.

This key opens the door to the solution of several other mysteries. For example, one can now begin to understand the origin of the extensive caverns in the circum-Mediterranean lands and the peculiar topography (called "karst") of Yugoslavia, where sinkholes and pinnacles abound. One can also provide an answer to the longstanding question of why ground-water circulation once penetrated 3,000 meters below sea level in a mid-ocean island such as Malta.

Not only the geomorphic changes associated with the desiccation of the Mediterranean but also the biological changes were catastrophic. Shortly after my return, Giuliano Ruggieri of the University of Palermo wrote me that he and his colleagues had long had faunal evidence that the Mediterranean must have gone through some radical changes in salinity during the late Miocene, when diverse marine lives died out and were replaced by a few species that could tolerate great variations in salinity.

The presence of a hot desert where the Mediterranean Sea is now should have had a serious climatic impact. Indeed,

European paleontologists have noticed that there was a change toward aridity in central Europe during the late Miocene, when the Vienna woods were turned into steppes. With the return of marine waters to the Mediterranean in the Pliocene Epoch, the central European climate became wet and cold and deteriorated gradually into the ice age.

The evolution of plants, tied closely to climatic changes, would thus have been dramatically affected by the desiccation of the Mediterranean. With increasing aridity, perennial plants might die out, to be replaced by a mutant strain of annuals whose seeds might remain dormant and survive long years of drought. Insects too might get rarer, and cross-pollinating plants depending upon the bees and other insects might then lose the battle to self-pollinators. Since the publication of our work, I have been contacted by specialists on Mediterranean floras who have told me, for example, that the genus *Medicago* seemed indeed to have evolved in such a fashion in response to the Messinian crisis. Species of oats also went through a similar path of evolution.

Even a partial desiccation of the isolated Mediterranean would have resulted in a lowering of the "sea level" within the basin, and thereby an increase in the relative elevations (above sea level) of circum-Mediterranean lands. This in turn should have led to a lowering of vegetational zones: low river plains now became high mountain meadows; newly exposed continental shelves became rims of plateaus where coniferous forests grew; while salt marshes were found near shallow briny lakes on abyssal plains. Such a floral distribution during the Messinian is exactly what has been recognized by a botanist colleague of mine at ETH, Gilbert Bocquet. Mediterranean islands, such as Corsica and Crete, were 4,000-meter mountain peaks during the Messinian desiccation, and the Alpine floras on those islands became isolated after the Pliocene deluge and became endemic floras now growing on island coasts, far below their usual altitudinal limits. In fact, Bocquet told me that the present distribution of the circum-Mediterranean floras had made no sense at all until many of the puzzles were resolved by a new floristic model based upon the assumption that the Mediterranean was a desiccated deep basin some 5 million years ago.

One is tempted to speculate that the increasing aridity and deforestation might have triggered the hominid evolution. After all, physical anthropologists used to tell us that monkeys were evolving into bipeds when they climbed down from the trees to search for food in the savannas. The idea is tantalizing, but we have no facts, except that the earliest ape-man might indeed be about 5 million years of age.

Vertebrate paleontologists did have facts that led them to suspect large-scale migration of land animals during the Miocene. African antelopes and horses could gallop to Spain across the isthmus of Gibraltar before it was split asunder to become the strait. African rodents could sneak from the south to build their new homes in Europe. And apparently, hippopotami made their way from the Nile to Cyprus. The migratory traffic might have been more frequent if the wanderers had not had to travel across a desert 2,000 to 3,000 meters below sea level.

The sudden isolation of the Mediterranean islands after the Pliocene deluge also might have driven the stranded population to inbreeding and endemism. One zoologist wrote me about this possibility, based on his studies of the lizards of the Adriatic islands. Another was convinced that the dwarf antelopes of Mallorca and Menorca owed their fate to the isolation of these Balearic Islands from the mainland during the last 5 million years.

We were surprised by other even more mysterious clues. We might not agree with H. G. Wells that swallows acquired their habit of flying directly over the Mediterranean in those days when it was a dry land. We were nevertheless intrigued by a recent finding that eels living in rivers draining into the Mediterranean do not join their European and American relatives in the traditional "breeding ground" for eels under the Sargasso Sea. The southern European eels alone choose to breed in the Mediterranean. Did they acquire this habit 6 million years ago, when they could not jump across the Gibraltar Falls? We cannot be certain, of course, but the fact that the Mediterranean dried up permits some unorthodox suggestions to solve problems in biological evolution.

The disappearance of this large inland sea was probably not a unique event in geological history. The existence of large saline deposits indicates that there might have been other desiccated oceans. The famous Zechstein salts of northern Europe may be the residue of an inland sea that dried up 250 million years ago. The giant salt and potash deposits of Alberta and Saskatchewan, some 350 million years old, may have had a similar origin. The discovery that a small ocean basin can be converted into a desert has led us to reexamine the whole problem of salt deposition. Geologists used to worry about the occurrence of oceanic salt deposits under the Gulf of Mexico, under the South Atlantic off the coast of the Congo and Angola, and under the North Atlantic off the coast of Nova Scotia. We can now postulate that these salts were formed when the Gulf and the Atlantic were isolated inland seas undergoing dessication.

It must seem somewhat farfetched to imagine the Mediterranean as a deep, dry, hot hellhole. It is thus not surprising that our interpretation has not been universally accepted even though the evidence permits no other alternative. Cita gets excited sometimes when her colleagues do not listen to her. But posterity will be the judge. A hundred and fifty years ago a young engineer from Geneva came up with the preposterous idea that central Europe had been covered by a giant ice sheet during the Pleistocene, because there was no alternative explanation for the presence of erratic boulders on the Swiss plateau. Nobody believed him. But his tormenters have long since passed away, and generations of schoolchildren have been told about the ice age; we all now take for granted that such an improbable event did indeed occur. The amount of water in the Mediterranean is of about the same order of magnitude as the bulk of ice that once covered northern Europe. To remove that much water from an ocean is just as improbable as to pile that much ice on land. But one of these days Cita's last tormenter will also meet his inevitable destiny, and new generations of schoolchildren will be taught to consider the desiccation of the Mediterranean a gospel truth.

POSTSCRIPT

—O—

LEG 13's discovery of an evaporative formation beneath the Mediterranean proved that salt exists under the deep-sea floor and that giant salt deposits can be formed within a relatively short span of geologic time. The almost synchronous onset and termination of the Mediterranean salinity crisis implies catastrophic changes of environment in a region over 2.5 million square kilometers in extent. Our interpretation that the evaporite formation was deposited during a time when an originally deep Mediterranean Sea was evaporated dry induced mixed reaction from the scientific community. Favorable commentaries were not rare, but many critiques were skeptical or downright hostile.

The idea is not difficult to stomach for those who are used to inductive and deductive reasoning. The two factors in question are the depth of the basin floor and the depth of the brine pool within the basin. If we use the two adjectives "deep" and "shallow" to describe the depth, combinations and permutations would give four alternatives: 1) deep basin, deep water; 2) shallow basin, shallow water; 3) deep basin, shallow water; and 4) shallow basin, deep water. Of the four, the last is physically impossible. And the geological data unearthed by the 1970 drilling expedition were sufficient to prove the impossibility of the first two. Using the maxim of Sherlock Holmes, what remains is the deep basin, shallow water, or the desiccated deep basin theory, however improbable it may seem. Unfortunately, the discovery made news, and the story got into newspaper headlines and popular journals long before the scientific report was published. Distorted by second-hand reporting that left out all technical details, our postulate of a desiccated Mediterranean provoked considerable irritation that may have retarded the acceptance of the theory by some. Many colleagues, engaged in investigations remote from the Mediterranean, simply dismissed the story with a wave of the hand. Others, having worked for many years on the problem, had developed strongly rooted prejudices and preferred to overlook our findings if they contradicted their pet ideas. Still others seemed to have had too narrow a training to understand some of the implications of the new discoveries. A few colleagues clung to their idea of deep brine precipitation, ignoring the evidence of the stromatolites, which indicated a water depth so shallow as to permit sunlight penetration for the growth of algal mat on the deep basin floor. Others continued to insist that there had been a shallow basin floor during the evaporite deposition, ignoring totally the fact that a basin underlaid by the oceanic crust had to be thousands of meters below sea level.

Although Ryan, Cita, and I were convinced of the validity of our postulate, we did seek an opportunity to look deeper into the question. We had arrived at a far-reaching conclusion on the basis of a modest investment of ship-time and a very modest amount of recovered materials. Our only solid evidence for the presence of evaporites in the eastern Mediterranean consisted of a few chips of gypsum in a mud that had been scraped off the drill bit after the drill string had been taken out of Hole 125. We had made no positive identification of potash salts. We had not penetrated the base of the evaporite formation to study the beginning of the salinity crisis, and we had very few cores to prove that the Mediterranean had been a deep sea before it was desiccated. We also had little data to tell us when (and how) the Mediterranean Sea had become neither a deep sea nor a salt desert, but a great lake. The evidence was so fragmentary that the Leg 13 scientists could not reach a concensus. The shipboard report presenting the desiccated deep basin theory was authored by only Maria Cita, Bill Ryan, and me, because we were the only ones convinced of its validity; the others preferred alternative explanations. We were therefore pleased when the JOIDES decided to schedule a second drilling cruise to the Mediterranean, after funding for the Deep Sea Drilling Project had been extended for a third time by the National Science Foundation through autumn of 1975.

The *Glomar Challenger* sailed on April 2, 1975, from Malaga to start Leg 42A of the Deep Sea Drilling Project, and returned to port in Istanbul on May 21. Lucien Montadert, of the Institut Français du Pétrole, and I were co-chief scientists, and we were assisted by ten shipboard scientists from France, Germany, Italy, Great Britain, Switzerland, and the United States. Maria Cita and I were the only two veterans of the Leg 13 cruise; all the others were newcomers selected by a JOIDES panel to insure scientific objectivity. During the thirty-seven days at sea, the *Glomar Challenger* traveled 6,000 kilometers, drilled eleven holes at eight sites, penetrated a total of 4,461.5 meters of sedimentary sequences, and recovered 670 meters of cores. Yet I could not help but feel a letdown; I missed the excitement of discovery experienced during the first cruise. The task of the second expedition was more somber and more difficult. But we had our share of joy, anxiety, and disappointment, and the technical achievements were not inconsiderable. We managed to do what we could not have done five years previously— namely, to penetrate the Mediterranean evaporites so as to obtain a record of the earlier Mediterranean history. We found unequivocal evidence that the Mediterranean had been a deep sea, for 15 million years at least, prior to the Messinian desiccation. We obtained many samples of evaporites in the eastern Mediterranean, as we had set out to do. We found potash salts at the location we predicted: these more soluble salts, last to crystallize out of a dense brine, should have been, and were, in the deepest center of a Mediterranean basin (Site 374, fig. 2). If we should ever develop a technique of deep-sea mining, the potash salt beneath the Mediterranean might provide an almost inexhaustible supply of chemical fertilizer.

We further established the fact that the Mediterranean had been a brackish lake for a few hundred thousand years after the salt deposition. The catastrophic desiccation apparently induced a reorganization of the European drainage system. Great rivers in eastern and central Europe had supplied fresh water to a giant lake that stretched from Austria to the Ural Mountains (fig. 39), but when the Mediterranean dried up, they were captured by the Mediterranean drainages. The influx of a vast quantity of fresh water into the salt desert gave rise to one or a series of great lakes (fig. 40).

I was pleased that the scientists on board reached a near-consensus in support of the theory of a deep desiccated basin during the Messinian salinity crisis. The final report of the

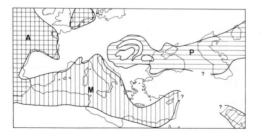

39. Disposition of the Paratethys (P) (the Black Sea, the Caspian Sea, and the Aral Sea are the modern remnants) and the Mediterranean (M) 15 million years ago.

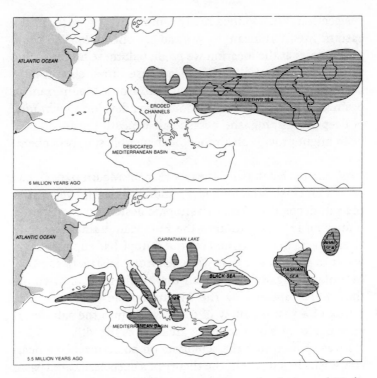

40. Waters from the Paratethys drained into the desiccated Mediterranean basins to form a series of brackish lakes there.

Leg 42A cruise represented the majority opinion of the shipboard scientists. There was no question about it: the Mediterranean had been in existence long before the salinity crisis, and the deep sea had dried up when the evaporites were formed.

GLOSSARY OF TECHNICAL
TERMS AND ACRONYMS

ABYSSAL PLAIN   a flat basin under the ocean, commonly greater
than 2,000 meters deep

AIR GUN   energy source for the CSP (q.v.)

ALBORAN BASIN   Mediterranean seabed between Spain and Mo-
rocco

ANHYDRITE   anhydrous calcium sulfate, $CaSO_4$; an evaporite (q.v.)

ARROYO   dry creek

BALEARIC BASIN   Mediterranean seabed between France and Algeria

BASALT   an igneous rock, commonly a congealed lava

BASEMENT   solid rock at the base of a sedimentary sequence, commonly a basalt in the ocean

BEACON, SONIC   a device that sends signals from the sea floor to
the drill ship to prevent her from drifting away from a drill hole

BIOSTRATIGRAPHY   use of fossils to study age relations of rock formations

BOTTOM-HOLE ASSEMBLY   the lowest part of a drill string, including
the drill bit, drill collar, and so on

BULL'S-EYE   referring to concentric distribution of saline minerals
in a desiccated basin, particularly its central part where rock salt
is often found

BUOY   a float marking the target area, while the drill ship is getting
set to stop

CALYPSO   French research vessel

CARBONATE ROCKS   limestones and/or dolomites (q.v.)

CENTER BIT   an auxiliary drill bit inside the drill collar

CHAIN   U.S. research vessel, operated by Woods Hole

CHARCOT   French research vessel

CHERT   a very hard rock composed of silica (SiO_2)

CNEXO   Centre National pour l'Exploitation des Océans

CONRAD   U.S. research vessel, operated by Lamont

CONTINENTAL SHELF   part of seabed above 200 meters

CONTINENTAL SLOPE   part of seabed extending from the outer edge of the continental shelf to the fringe of the abyssal plain (or trenches)

CORE BARREL   a steel, open-ended barrel that is lowered inside the drill string before a core is taken

CORE CATCHER   a device at the end of a core barrel to prevent soft core sample from dropping out while the barrel is being hauled up

CORING   act of taking a core from a drill hole

CORING, CONTINUOUS   to obtain a complete set of core samples from a drill hole

CORE LAB   laboratory for sedimentologists and marine technicians on *Glomar Challenger*

CRETACEOUS   a period in earth history between 130 and 65 million years ago

CSP (continuous seismic profiling)   investigation of seabeds by registering travel times of artificially produced acoustic signals

DEFORMATION   referring to disturbance of rock formation, which might be contorted or fractured or displaced

DETRITAL   pertaining to detritus, or debris

DIATOM   a siliceous microorganism; marine brackish or freshwater

DOLOMITE   a rock composed of the double salt $CaMg(CO_3)_2$; certain dolomites are evaporites (*q.v.*)

DRILL BIT   tool for digging at the lower end of a drill string

DRILL COLLAR   the lowest pipe of a drill string, to which the drill bit is attached

DRILLER'S SHACK   a partially enclosed place where a driller operates his tools to drive the drill string

DRILL PIPE   steel pipe segments, 9 (single), 18 (double), or 27 (triple) meters long, which can be joined together to make a drill string (*q.v.*)

DRILL STRING   a string of drill pipes with the bottom-hole assembly (*q.v.*), at its lower end; the string connects the ship to the drill hole

DSDP   Deep Sea Drilling Project

"EELS" a string of instruments used in CSP (q.v.), towed by a research vessel

EVAPORITE precipitate from a natural saline solution consequent upon evaporation

FIX short for sat-fix

FORAMINIFERA one-cell animals; their skeletons are common in ocean sediments

GABBRO a dark igneous rock, an ophiolite (q.v.)

GEOSYNCLINE elongate depressions on earth's surface where thick sediments were accumulated

GNEISS a metamorphic rock (q.v.)

GRADED BED a sedimentary layer showing an upward decrease in size of grain

GRANITE an igneous rock common on continent

GYPSUM hydrous calcium sulfate, $CaSO_4 \cdot 2H_2O$; an evaporite

ICE AGE see Pleistocene

IONIAN BASIN Mediterranean seabed south of Greece

IPOD International Phase of Ocean Drilling

JOIDES Joint Oceanographic Institutions for Deep Earth Sampling

JOIDES, EUROPEAN FRIENDS OF an informal group of European scientists, interested in JOIDES drilling

LA JOLLA referring to Scripps

LAMONT Lamont-Doherty Geological Observatory of the Columbia University, New York

LEG 3, DSDP a drilling cruise to the South Atlantic, 1968-69

LEG 6, DSDP a drilling cruise to the Pacific, 1969

LEG 10, DSDP a drilling cruise to Gulf of Mexico, 1970

LEG 13, DSDP drilling cruise to the Mediterranean, 1970

LEG 42, DSDP second drilling cruise to the Mediterranean, 1975

LITHIFIED changed into rock

"MAGGIE" instrument towed by a vessel to measure magnetic properties of seabed

MARL a sediment composed of clay minerals and calcium carbonate

MÉLANGE mixture of different rock types formed by mountain building processes

MERCAST communicating messages through a form of broadcasting

MESOZOIC an era in earth history between 220 and 65 million years ago, including three periods: Triassic (q.v.), Jurassic, and Cretaceous (q.v.)

MESSINIAN   a stage in earth history between 6 and 5 million years ago, when the Mediterranean was repeatedly desiccated

METAMORPHIC ROCK   a rock that has been altered from another, commonly consequent upon deep burial

MIAMI   referring to the Rosenstiel School of Marine and Atmospheric Sciences, University of Miami

MIOCENE   an epoch in earth history between 25 and 5 million years ago

MOHOLE   an unsuccessful attempt to drill beneath the earth's crust

M-LAYER   layer of sediments beneath the M-reflector

M-REFLECTOR   top of the Mediterranean evaporite

NANNOPLANKTON   minute pelagic (q.v.) plants; their calcareous skeletons commonly constitute the bulk of ocean sediments

OFFSET HOLE   a hole drilled near the first hole of a drill site after the vessel has been moved a short distance; the drill string is pulled out of the original hole but not hauled up on deck

OOZE   wet ocean mud, composed largely of foraminiferal and nannoplanktonic skeletons

OPHIOLITES   rocks forming ancient ocean floor

OSTRACOD   small molluisklike animal, marine brackish or freshwater

PALEO LAB   laboratory where paleontologists work on *Glomar Challenger*

PDR (PRECISION DEPTH RECORDER)   a precise echo-sounder

PELAGIC   of the open sea

"PILLAR OF ATLANTIS"   referring to core of anhydrite

PISTON CORE   a core taken by conventional research vessel, commonly shorter than 20 meters

PLAYA   flat floor of desert basin

PLEISTOCENE   an epoch in earth history between 2 million and 10,000 years ago; ice sheets covered large parts of northern continents repeatedly during this epoch

PLIOCENE   an epoch in earth history between 5 and 2 million years ago

QUATERNARY   the fourth and last era of earth history, starting 2 million years ago

QUARTZITE   a rock composed of quartz

RADIOLARIA   one-celled marine organisms with SiO_2 test

REFLECTOR   layer under sea floor capable of bouncing back acoustic signals

RHYOLITE   a volcanic rock

ROUGHNECKS   hardest workers on *Glomar Challenger*

SABKHA   arid salt marsh or coastal flat

SALINITY CRISIS   a crisis for marine organisms when ocean salinity becomes intolerably abnormal

SALINITY CRISIS, THE   late Miocene crisis in the Mediterranean

SALT DOMES   pillar-shaped bodies of salt, extruded into a sedimentary sequence

SANDLINE   a steel rope lowered inside the drill string to fish out a core barrel (q.v.)

SAT-FIX   the exact location of the ship determined by sat-nav (q.v.)

SAT-NAV   satellite navigation

SCHIST   a metamorphic rock (q.v.)

SCIENCE OFFICE   room where the sat-nav computer sits, and where a typewriter can be found

SCRIPPS   Scripps Institution of Oceanography, La Jolla, California

SEISMIC PROFILER   see CPR

SERPENTINITE   an ophiolite (q.v.)

SHALE   a lithified (q.v.) mud

SHEAR PIN   a device at the end of a sandline (q.v.), designed to be hooked onto a core barrel, so that the barrel can be hauled up

SILICEOUS   pertaining to the chemical compound silica (SiO_2)

SPUD   begin digging

STRING   see drill string

STRUCTURES   referring to the geometry of deformed rocks (see deformation)

STROMATOLITE   calcareous rock, commonly containing algae

TEPHRA   volcanic ash

TETHYS   a postulated ancient seaway between Eurasia and Africa, named after a goddess in Greek mythology

TOOLPUSHER   a supervisor of drilling operations

TRUBI   name of a Pliocene (q.v.) formation directly overlying the evaporite in Sicily

TURBIDITY CURRENT   an underwater current full of mud and gravel suspensions

TRIASSIC   a period in earth history between 220 and 170 million years ago

U.S.S. BUTTES   a vessel from the U.S. 6th Fleet

VALENCIA TROUGH   Mediterranean seabed between Spain and Balearic Islands

WOODS HOLE   The Woods Hole Institution of Oceanography, Woods Hole, Massachusetts

YEOMAN   euphemism for secretary on *Glomar Challenger*

SUGGESTIONS FOR
FURTHER READING

Numerous books and technical articles have been published on the desiccation of the Mediterranean. Of those, I suggest the following for those who are interested in the technical details:

Initial Reports of the Deep Sea Drilling Project, vol. 13, ed. W.B.F. Ryan and K. J. Hsü, and vol. 42A, ed. K. J. Hsü and L. Montadert. Washington, D.C.: U.S. Government Printing Office, 1973 and 1978.

K. J. Hsü, W.B.F. Ryan, and M. B. Cita. Late Miocene Desiccation of the Mediterranean. In *Nature* 242 (1973):240-244.

K. J. Hsü, L. Montadert, D. Bernoulli, M. B. Cita, A. Erickson, R. E. Garrison, R. B. Kidd, F. Mèlierés, C. Müller, and R. Wright. History of the Mediterranean Salinity Crisis. In *Nature* 267 (1977):399-403.

M. B. Cita and R. Wright, eds. Geodynamic and biodynamic effects of the Messinian salinity crisis in the Mediterranean. *Palaeogeography, Palaeoclimatology, Palaeoecology,* 29, nos. 1-2 (1979).

INDEX

—Θ—

Library of Congress Cataloging in Publication Data

Hsü, Kenneth J. (Kenneth Jinghwa), 1929-
The Mediterranean was a desert.

Bibliography: p.
Includes index.
1. Geology—Mediterranean Sea. I. Title

QE350.22.M42H78   1983   551.46'2   83-2269
ISBN 0-691-08293-6
ISBN 0-691-02406-5 (pbk.)